SOILI TURUNEN

Lilly Kolisko - The Mystery of Matter

SOILI TURUNEN

Lilly Kolisko
The Mystery of Matter

Documentary | Biography

ISBN 978-1-9191836-0-2

Contents

Foreword

L illy Kolisko (1889 - 1976), born Elisabeth Anna Noha, was an outstanding scientist. She succeeded in developing the anthroposophical approach to a "science of life that places the human being at the centre" through decades of meticulous and relentless empirical research. She carried out her experiments, especially her rising picture experiments, under difficult and sometimes adventurous conditions; during a total solar eclipse in 1936 in her hotel room in Turkey, on a sea voyage to India, or down a 16-metre-deep shaft in Stuttgart. This had been excavated especially for her research into the seasonally varying crystallisation forces and the influence of the moon on plant growth. Whilst the external circumstances of her work might appear modest to primitive, her approach and her results were qualitatively convincing. Scientific papers are published based on Lilly Kolisko's work and methods even today.

Lilly Kolisko accepted invitations to give public lectures often using colour slides to illustrate her unusual research work and results. She could provoke astonishment like at the Vienna Society of Engineers and Architects in 1928 as well as before Indian doctors and scientists in Madras (1937) and again in London, with newspaper reports in "The Daily Mail" and the "Observer" among others. She addressed chemists, physicists and technicians in a machine factory in Stuttgart-Ober Türkheim. In India, she not only held extensive discussions with homeopathic doctors and scientists but also met with the philosopher Rabindranath Tagore. Together they explored ancient, traditional folk medicine in harmony with the latest research on the "effect of the stars upon earthly matter."

Sensitive people who heard her explanations and studied her publications were deeply impressed by her downright "cosmic certainty" (F. Geuter) and saw her as a scientist of the future. The Englishwoman Eleanor Merry, herself a prominent figure, spoke of Lilly Kolisko's "unique genius ... I have never met a woman [like Lilly Kolisko] with such undaunted faith and courage" (E. Merry). According to Adalbert von Keyserlingk, senior pupils at the Stuttgart Waldorf School who received science lessons from Lilly Kolisko and followed some of her experiments (including in the deep shaft on the school grounds), were "breathlessly interested".

After graduating from high school, she intended to study medicine and become a doctor. However, she came from a poor background, had to earn money and so worked as a secretary using her facility with foreign languages.

In the First World War her destiny led her to learn laboratory techniques in a Viennese hospital where she met the medical student Eugen Kolisko. Years later, between 1921 and 1923, she was enrolled at the universities of Tübingen and Stuttgart for medicine and natural sciences. However, due to her family and professional situation - mother to her daughter Geni and already in the middle of experimental research work - she was unable to fully pursue and complete her studies.

Lilly Kolisko began her scientific work in Stuttgart in the early 1920s. This centred on physiological, pathophysiological and veterinary issues (L. Kolisko: "Function of the Spleen and Blood platelets"; E. Kolisko: "The Nature and Treatment of Foot-and-Mouth Disease") included studies on the efficacy of homoeopathic preparations which Rudolf Steiner found "brilliantly" successful (L. Kolisko: "Physiological and Physical Proof of the Efficacy of the Smallest Entities"). This work continued to develop brilliantly after the 'agricultural course' in Koberwitz Castle (Silesia, June 1924). "All scientific research work [on the foundations of the new agriculture] must be carried out by you," Steiner said to her in Koberwitz.

Lilly Kolisko published "Agriculture of Tomorrow" in English in 1946. This contained the results from 22 years of her own work and that of joint studies with Eugen Kolisko.

Lilly Kolisko always emphasised that she did not carry out commissioned or confirmatory research but worked "in complete individual freedom". This was indeed the case: "It was an honest pursuit of truth and knowledge." She received the answers from her experiments - which focussed on her method of "capillary dynamolysis" - not from Rudolf Steiner, but from the cosmos. Her results emerged from the matter itself, from the forces, processes and substances investigated in space and time. How much Rudolf Steiner valued Lilly Kolisko's work for the scientific impulse of the Goetheanum can be seen from her presentation at the Christmas Conference 1923/24 (on 31 December 1923) and Steiner's comments: "But all these experiments are basically details of a totality that is actually needed as urgently as possible in science today, especially from the anthroposophical point of view."

In this context, Steiner spoke with Ita Wegman and Eugen Kolisko about the incorporation of Lilly Kolisko's work into the Goetheanum ("Biological Institute at the Goetheanum", based in Stuttgart), which was subsequently agreed.

After the esoteric classes at the School of Spiritual Science began on 15 February 1924, Lilly Kolisko was in Dornach weekly, taking notes of the lessons and, at Rudolf Steiner's request, giving them to the faculty of the Stuttgart Waldorf School.

She understood that her scientific work was closely linked to spiritual training - Rudolf Steiner asserted this for the entire school with all its departments or sections. On his last birthday on earth on 27 February 1925, she wrote to him about the results of her research since the Koberwitz course; she observed the wake for Dr Steiner five weeks later for one night, together with Ita Wegman and Count Keyserlingk of Koberwitz.

In 1928 she spoke at the opening celebrations of the Second Goetheanum about her research into the annual cycle as reflected in substances (metal salts) with her familiar colourful slides. She was able to demonstrate the Christian annual festivals as revealed in the change of substances, thus visualising the reality of Steiner's imaginations of the seasons in a scientific way. She had been working tirelessly on this research for years including throughout the Holy Nights. Her sentence "The whole universe is pervaded by the forces of resurrection" did not stem from faith, but from scientific experience. At the opening of the Second Goetheanum in Michaelmas 1928, she brought with her what she considered to be the most valuable aspect of her research to date.

However, Lilly Kolisko's life was also marked by a bitter trace of tragedy. The disputes in the Anthroposophical Society after Rudolf Steiner's death led to a deep rupture of the anthroposophical community. Eugen Kolisko was expelled from the Anthroposophical Society, Ita Wegman was stripped of all her offices (1935) - and much came to an end. It was not until 1957 to 1960 that Lilly Kolisko found the time and opportunity to write a biography of her husband in order to come to terms with the distortions and alienations of which he was not the only victim.

Her life was overshadowed by Eugen's early death (from a heart attack in November 1939 whilst in exile in England), by National Socialism, and by the Second World War and other intervening events. Nevertheless, even after Eugen Kolisko's death, she continued her high-quality work in England with a scientific dedication and loyalty that could hardly be surpassed.

During her visits to Germany after 1949, she impressed the Weleda specialists and attendees at her lectures with her expertise, and the members of the School of Spiritual Science with the spiritual diction and substance of the classes she gave. Until the very end, she found her research – "my endless work - incredibly interesting". One of her poems says: "never stop growing within yourself."

*

Gunhild Pörksen and I at the Ita Wegman Archive and Institute have followed Soili Turunen's conscientious and careful studies on this documentary-biography of Lilly Kolisko with great interest from the very beginning. Now the author has succeeded in completing the German edition and bringing it to publication for the 100th anniversary of the Koberwitz Course and the direction of work taken there, with which Lilly Kolisko (like Elisabeth Vreede and Ita Wegman) was closely associated. We are delighted about this!

The Ita Wegman Archive and Institute would like to thank Soili Turunen and all those who accompanied her on this long journey and supported the publication of the book, first and foremost Gerrit Overweg. What Lilly Kolisko achieved for Eugen Kolisko and his rehabilitation with her biography, in the portrayal of his actions and motivations in a difficult historical environment, Soili Turunen has now achieved for Lilly Kolisko and her life's work.

"Long is the time, but what is true comes to pass..."
(Hölderlin: *Mnemosyne IV*)

Peter Selg

General Anthroposophical Section (Goetheanum)
Ita Wegman Institute (Arlesheim)
June 2024

Acknowledgements

I n 1983, the name *Lilly Kolisko* first came to my notice when I was working in the "Nordisk Forskningsring" (Nordic Research Circle for Biodynamic Research) with the copper chloride crystallisation method under the direction of Bo Petterson and delving into the rising picture method with the help of Magda Engquist. Lilly Kolisko's book, "The Agriculture of Tomorrow", made a great impression on me. One day, Uwe Lemcke, a priest in the Christian Community, came by the lab and asked if I was familiar with Lilly Kolisko's work on the annual festivals and if I could repeat it. He later brought me a copy of the book "Spirit in Matter". The photocopying machines of the time could only reproduce the pictures very poorly, so I could only imagine how the original pictures would look with the help of Lilly Kolisko's descriptions.

I began to enquire wherever I could about Lilly Kolisko, her work and her life. The answers came that although there were some publications about her research, they were very difficult to find in antiquarian bookshops. I also heard that Lilly Kolisko had burnt her research materials before she died. Very little could be found about her life.

Later I took part in the "Arbeitskreis Bildschaffende Methoden". There Janet Barker, in particular, helped me by introducing several contacts in England and obtaining many of the publications by Lilly Kolisko that I was missing. Heidi Pflückiger gave me a scarf handmade by Lilly Kolisko that once belonged to Agnes Fyfe.

My life path then led me to the Camphill community in Norway, and I told my Camphill contacts in England and Scotland - especially Friedwart Bock - that I was trying to track down the original materials for the capillary dynamolysis pictures. In Buckfastleigh, England, I met Cecil Reilly, who told me that after Lilly Kolisko's death he received a parcel from a nurse who had cared for his wife when she was very ill, but also for Lilly Kolisko in her last years. He said, "destiny is speaking", and gave me all the materials he had received, including about 60 original capillary dynamolysis pictures and several letters from Lilly Kolisko. I later learnt that this nurse was Ilse Ketzel, who lived in Tuffley, Gloucester, UK in the house where Norbert Glas had had his medical practice and had provided Lilly Kolisko with some study rooms. After Lilly Kolisko's death, Ilse Ketzel also gave other people various research materials, including slides that Lilly Kolisko had shown during her lectures in later years.

I met several times with Andrew Clunies-Ross, the grandson of Lilly and Eugen Kolisko, who took over the "Kolisko Archive" after Lilly Kolisko's death and edited further publications and published reprints of works by Eugen and Lilly Kolisko. He kindly showed me the research materials from the Stuttgart period, various letters and stenographic notes, and he also kindly gave me permission to publish some photos from the Kolisko Archive. He offers many of the books in digitised form (www.koliskoarchive.com).

A very important place for research and support for this biography was the Ita Wegman Archive, where I received a very warm and open welcome from Peter Selg, Gunhild Pörksen and Felicitas Graf. Finally, Mirela Faldey, who also works in the Ita Wegman Archive, helped me with the necessary documents and images.

The correspondence between Ita Wegman and Lilly Kolisko gives a vivid picture of the years 1922 to 1939. Other letters from the Ita Wegman Archive were also important for the biographical description.

The Goetheanum Archives and the Rudolf Steiner Archives also contributed important documents.

Anne Weise, who did a valuable job digitising the Karl König Archive, was able to unearth important documents using the search term "Kolisko", especially from the period after the Second World War.

During my archive searches, I also met Christoph Podak who was generous in sharing his findings with me.

The Finnish foundation "Signe ja Ane Gyllenberg Säätiö" supported my project with research funds, for which I would like to express my sincere thanks!

I would like to thank the following people in particular: Markus Schultze-Florey for the first proofreading of the German edition down to the smallest punctuation and commas, Brigitte Kowarik for the very important linguistic improvements and Mats Miersch, Clara Grundmann and Siri Spence for the translation of letters and documents from English. Natalia and Sven Baumann did the proofreading in a very nice way.

During many holiday trips "in the footsteps of Lilly Kolisko", my life companion, Gerrit Overweg, was an important help.

I am aware that the beginning and end of Lilly Kolisko's biography are still particularly incomplete. I am also sure that many of her research materials and letters are still stored somewhere.

So, a heartfelt request: It would be very important to bring together all the surviving materials in one place so that future researchers can find, study and further develop Lilly Kolisko's important research approaches. Andrew Clunies-Ross has made a start by handing over a large part of the Kolisko archive to Thomas Meyer (Perseus Verlag), Arlesheim.

And many thanks to all those who supported the publication of the German edition of this book with donations!

Soili Turunen
soili.turunen@gmail.com
Trondheim (Norway), February 2024

Following the publication of the German edition of this biography, a question arose regarding an English translation. Mark Moodie's warm interest for Lilly Kolisko and her work made this possible, and I would like to express my sincere thanks to him for his courage in taking on the task of editing and publishing this comprehensive biography.

I would also like to thank Corinna Bruckner Balavoine; her corrections and advice were essential in improving the translation. Please note that some of the letters reproduced here were written in English by persons who had German as their mother tongue and are kept in their original version.

I hope that Lilly Kolisko's life and work can provide new insights and inspiration for all English-speaking readers around the world.

Soili Turunen
Trondheim, November 2025

1.

Early years

I still can't believe that so much love has been given to me. I have received so little love in my life and basically, I walked amongst people as a stranger. I wanted to give people everything, but I kept them away from my inner self, from my soul. The many sufferings had also filled me with unspeakable bitterness ...[1]

There are few traces of Elisabeth Anna Noha's life before she moved to Stuttgart. Even her exact date of birth is unclear. Her last English passport gives her date of birth as 3 September 1889. Gisbert Husemann suggests 1 September in his memoirs of Lilly Kolisko.[2] Eugen Kolisko, however, repeatedly mentions 31 August as "Lilly's birthday".[3] This was supported by Gladys Knapp, a close friend of Lilly's from England: "She was born in Vienna on 31 August 1889."[4] The entry in the birth register in Vienna lists her birthday as 1 September 1889, but we must take into account that the custom common in Austria at the time was to use the date of registration in baptismal or birth register as the birth day.

Elisabeth Anna was the daughter of Leopold Noha (1855 - 1935) and his second wife, Anna Barbara (née Groß). She had two half-brothers from her father's first marriage. The family first lived in Vienna's Gürtelstraße and from 1905 in Hernalser Gürtel 15. Her father was a typesetter by profession.[5]

Gisbert Husemann wrote in his obituary of Lilly Kolisko in 1978: "Her father was a typesetter. She had two step siblings. Apart from poverty, her home life was marked by her father's tendency to drown his troubles in drink. When he came home in the evening Lilly's mother would send her to him to calm him down."[6]

In 1937 Lilly Kolisko was living in England. In the journal "Modern Mystic", she recalled a childhood incident that gave the impression of a special child who overcame her fear through interest and observation:

1

I remember one evening, in my fourth or fifth year, being very naughty. My mother told me to cease immediately, or something would happen. I simply could not stop. Not only did I wish to keep on, but I was now intrigued to know what it was that would happen. So, I was sent to a dark room, my mother saying: "Look, there is the moon! He will come and fetch you because you are so naughty." Left alone, I trembled with fear but did not cry. With my back to the window, I saw the shadow of the curtains moving to and fro in the moonlight which fell on the opposite wall. After a time, drawn by some magnetic power, I walked to the window, pushed back the thin curtains, and looked up at the moon. It was almost full. I looked at it more and more intently, trying to see the "man" mirrored in its face, but I could not. Instead, I saw two angels, flying from its sides into the middle - an image which has remained - so that even now when I look at the moon I see them. Much later my mother came in and wondered why I was not crying. All my fear had gone. The moon had not "fetched" me; I only felt very tired. Ever since that day I have liked the moon![7]

Another incident from her early youth also shows that she was very alert to the phenomena of nature and searched for answers to phenomena that she did not know or understand:

I was about twelve years of age when, one Sunday in mid-summer, my father took me into the beautiful woods around Vienna. It became late and we were overtaken by darkness. At last we came to an open place and saw in the far distance the steeples of the town. The night was wonderfully clear and the sky full of twinkling stars. Suddenly there appeared on the horizon a bright, reddish-yellow light. It grew larger and larger, then a huge red ball appeared. I was highly excited and asked my father what it could possibly be. He hesitated a moment before saying, "Oh, I think it must be the moon." "No, that is impossible," I cried, "the moon never looks so red, and is much smaller." We still had a long walk before us, and my eyes remained directed to the red ball on the horizon. It rose higher and higher, the colour slowly changing to yellow. After a time, the orb became smaller, then

2

streamed out its silvery light, and was my old friend, the moon.

My father could not explain why the moon looked so enormous and had so red a colour. I was very much puzzled about this strange phenomenon and went on thinking about it. [8]

One explanation could have been a total lunar eclipse visible in Vienna o 22. April 1902 at 7.52 pm in Vienna, even if Lilly Kolisko stated the time in midsummer.

Another experience from her youth can be found in an article in the magazine "Natura", published by Ita Wegman:

Whenever autumn approaches an early experience stands vividly before my soul. It was a Sunday in autumn. I was standing at the window of my room, from which I could see a small garden with many trees. I was seized with such infinite sadness at the dying nature, at the yellowing of the leaves, at the withering flowers, the departing birds, the strange air; everything, everything was sinking into the earth, everything was dying, everything was becoming desolate. I felt so lonely, so abandoned. I carried the death of all of nature in me in this hour. And yet, when spring comes, trees will bud again, the flowers will bloom again, the birds will sing again. Every year they rise again, and man, when he dies - does he not rise again alone? Does he sink back into the earth and become earth itself without resurrection? Can the human spirit also die? Fear joined the endless anxiety and sadness. This autumn Sunday became a question of life for me. Never again have I experienced autumn to this extent, but every time autumn arrives, I remember this early experience.

What is going on in nature, in the stones, in the plants, in animals and people at this time of year? I would never have found the answer if destiny had not led me to anthroposophy [...].[9]

Fig. 1: Elisabeth Anna Noha. © Kolisko Archive, London

Fig. 2: Elisabeth Anna Noha with her parents. © Kolisko Archive, London

She found little understanding from her family and so, as a young woman, she began to search for the answers to her unanswered questions herself. She read books on philosophy, astronomy, meteorology and other scientific works. She never told anyone about her great desire to become a doctor, "[...] to develop a medicine for a certain disease"[10.]

Elisabeth Anna Noha completed her schooling with the Matura. A short anecdote about the Matura exam shows how much more she had learnt than was required. She proposed Hebbel as her chosen author for the exam and when the professor said that he was not prepared for it, she replied: "It's enough if I'm prepared."[11]

It was financially impossible for her to continue her studies at university. Elisabeth Anna Noha worked as a secretary in several Viennese companies, where she corresponded in various languages (German, English, French and Italian).

Fig. 3: Rudolf Steiner, 1915
© Documentation at the Goetheanum, Dornach

2.

Important encounters

"You see the ether."[12]

The outbreak of the First World War also changed the destiny of the now 15-year-old Elisabeth Anna Noha. She volunteered at a military hospital in Vienna and initially helped to nurse the wounded soldiers. This is also where she met her future husband.

Lilly Kolisko later described this encounter to her friend, fellow helper and supporter Gladys Knapp:

> When war broke out in 1914, she volunteered for nursing, "on the battlefield if possible". She was told to report to a hospital in Vienna. She said: "All the other volunteers seemed to be elderly Viennese society ladies. A professor gave us a lecture on first aid and showed us how to lift a badly injured patient. Two medical students were there to be used by the professor in his demonstrations. One soon disappeared, but the other was so gentle and unassuming and patiently endured the ordeal of being lifted again and again. When my turn came, I lifted his head; he was Eugen Kolisko.[13]

Eugen Kolisko (21 March 1893, Vienna - 29 November 1939, London) was a medical student in the same military hospital. As the son of a renowned professor of forensic medicine and of a musically gifted mother from an aristocratic family, he belonged to the Viennese upper middle class. His education was deeply rooted in the science and art of the time.

Due to numerous operations in his childhood, Eugen Kolisko had to be privately educated and only entered the historic Schottengymnasium grammar school in Vienna in 1903. From 1911 onwards, he was a student at the medical faculty of Vienna University. Through his friendship with Walter Johannes Stein, he learnt about anthroposophy and became a member of the Anthroposophical Society shortly before his 21st birthday.

8

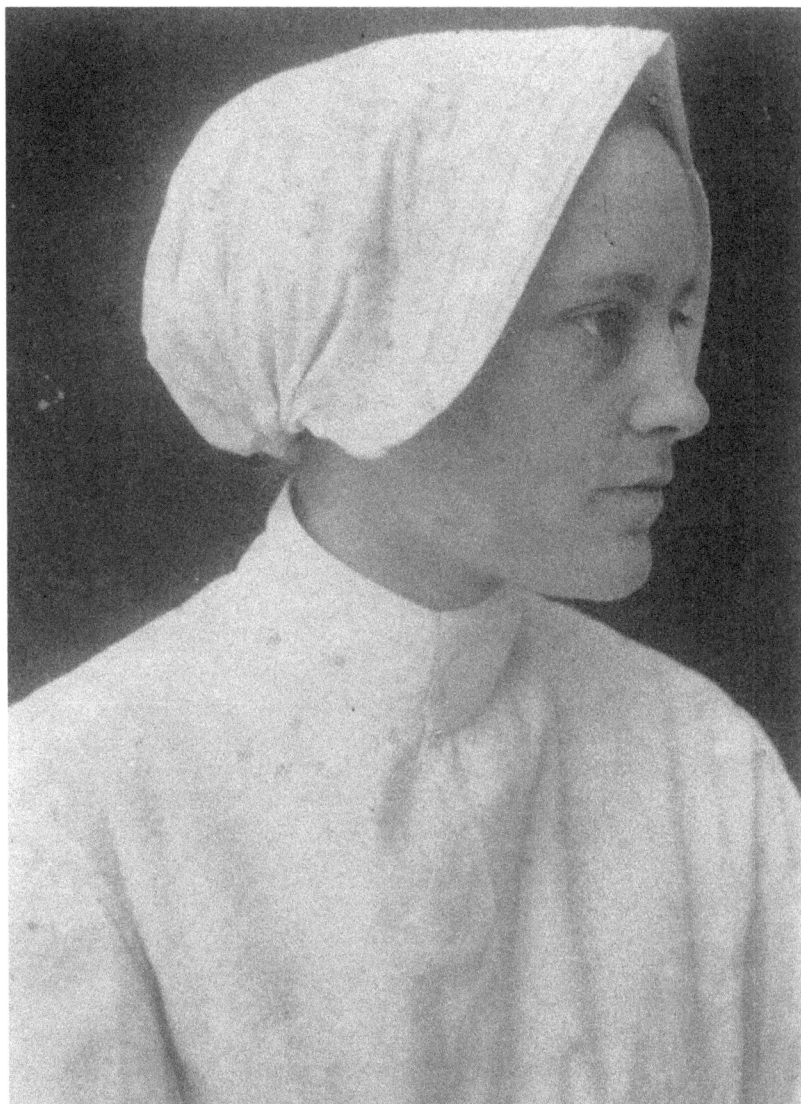

Fig. 4: Elisabeth Anna Noha as a nurse. © Kolisko Archive, London

Elisabeth Anna Noha learnt about Rudolf Steiner and anthroposophy for the first time in a conversation with Eugen Kolisko. Eugen Kolisko owned several of Rudolf Steiner's books, and so a new world opened up for her. She borrowed the book "How to Know Higher Worlds" and read it in one night. This book convinced her: Here was real knowledge, real truth. The exercises described by Rudolf Steiner became essential for the rest of her life. She later told Gladys Knapp in England: "If you neglect the exercise for just one day, you have interrupted the rhythm and have to do everything all over again [...]." And: "If you're waiting at a bus stop, don't throw away your time - it's a good opportunity to meditate!"[14]

From then on, she read all of Rudolf Steiner's books and lectures with great interest.

In May 1915, during a lecture by Rudolf Steiner at the Vienna branch of the Anthroposophical Society, she had a special encounter that was decisive for the rest of her life. She was "recognised" by Rudolf Steiner:

> Dr Steiner said "Yes, we've met before." She replied: "No." And she insisted on this "no", even though Rudolf Steiner repeated his acquaintance with her twice more until he finally rebuked her: "Yes, yes, from before." Eugen Kolisko, whose dismay one can only imagine, said to her standing there: "Don't you want to talk to Dr Steiner?" And again she said no. He advised her to write a letter to Rudolf Steiner. That's what she did. [...] At the next lecture, Dr Steiner approached her from the crowd: "You wanted to talk to me." - "Yes, I wanted the answer to my letter." Then Dr Rudolf Steiner said: "You have been through a lot of hard times and you don't sleep much." Then he advised her to imagine herself lying on a precipice, letting rose petals fall down and scooping them up again. Referring to her letter, he continued: "And then you asked about occult chemistry; you should fill in the blanks first, and only then deal with it!" And then he said directly to Lilly Kolisko: "You see the ether." It sounded like a riddle to her ears.[15]

Soon afterwards, on 15 November 1915, she became a member of the Anthroposophical Society.

During the war, blood and urine analyses were needed in hospitals for soldiers suffering from malaria, among other illnesses. Elisabeth Anna Noha volunteered for this laboratory work and became familiar with various methods: the cultivation of bacterial cultures, cell differentiation and the staining of blood smears.[16] This formed the basis for her later research work.

From 1915, during his final years of study, Eugen Kolisko was employed as an assistant at the Institute for Applied Medicinal Chemistry at the University of Vienna and worked there until March 1920. In July 1917 he received his doctorate in medicine.

On 5 December, 1917, Elisabeth Anna Noha and Eugen Kolisko married [17]. After the wedding, she changed her name to L. (Lilly) Kolisko.[18] The reason she gave was, "to have different initials from my husband".[19]

Eugen's parents had a hard time with the difference in status - the traditions of the old order were still too strong. His father only became reconciled when he was terminally ill. His mother never did, even though Eugen was her only surviving child after the death of his older brother Fritz.

The First World War ended on 11 November 1918. On 10 April 1919 Eugenie, Lilly and Eugen's daughter, was born.

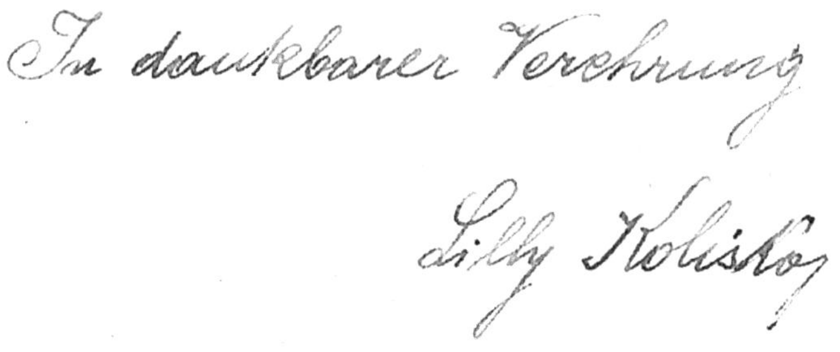

Fig. 5: Signature of Lilly Kolisko. © Rudolf Steiner Archive

Fig. 6: Lilly and Eugen Kolisko.
From the estate of Cecil Reilly / S. T. Norway

Fig. 7: Lilly Kolisko with Eugenie. © *Kolisko Archive, London*

3.

Intensive years - the beginnings of anthroposophical scientific research in Stuttgart from 1920

Guided by Dr Steiner's wise and kind hand, it was possible to carry out research in full individual freedom as a natural scientist in the anthroposophical sense. He paved the way everywhere. We just have to follow it.[20]

In the autumn of 1919, the first Waldorf School was founded in Stuttgart. In March 1920, Eugen Kolisko left his secure life and employment at the University of Vienna and moved to Stuttgart, initially to take on a position at the Waldorf School as a replacement for Friedrich Oehlschlegel. He was able to take part in Rudolf Steiner's scientific and educational lecture courses and give talks himself.[21] Eugen Kolisko then taught mainly chemistry, anthropology and hygiene. From October 1921 he also became the school doctor for the entire Waldorf School and blessed the school with his valuable contributions.[22]

For the time being, Lilly remained in Vienna with her daughter receiving many letters and reports about the events in Stuttgart and in Dornach. As early as June 1920, Eugen was asked by Rudolf Steiner to take on a new task:

Next week I will be starting a somewhat adventurous expedition. Dr Steiner indicated a possible treatment against foot-and-mouth disease. It is rampant here. "*Der Kommende Tag*"[23] owns several large estates, one of which is managed by Dr Rudolf Meier, who founded the research institute here and was formerly an assistant at the technical college. The contaminated barns are nearby. This is where the experiment is to be conducted, and I am to do this with Dr Steiner's consent and under

15

his instruction. A success would be of great importance. So I will go there for about eight days. For these few days I will be represented by Mr Stockmeyer.[24]

In March 1920, Lilly Kolisko also made an important inner decision and left the Catholic Church. In September 1920, she moved to Stuttgart with her daughter. The small family got a flat in the newly built teachers' residence.

She followed the scientific work in Stuttgart with great interest. The very severe outbreak of foot-and-mouth disease in Württemberg raised questions about new remedies and their production:

> Dr Steiner suggested a remedy, but this first required detailed testing. The correct and effective preparation of the remedy was particularly difficult. Frau Dibbern, in particular, endeavoured to achieve this. In the course of time, it was possible to prepare the remedy correctly, but this could only be achieved somewhat instinctively. Of course, such means are not always reliable. We wanted a more reliable criterion by which we could recognise whether the preparation was optimal. Again, Dr Steiner was asked how one could tell with certainty that the procedure had been successful. His answer was that this was *easily* possible. One should study the cell structure of the original substance and the prepared version and then one would find that a change in the structure of the protoplasm had occurred in the remedy.[25]

This question became Lilly Kolisko's first research task, and she was given a small room in the school's administration building for the purpose:

> This brings us to the origins of the Biological Institute. Dr Rudolf Maier had tried for some time to determine these structural changes using thin sections but had not achieved his goal. I found out about this and suggested making microscopic sections that could then be stained in the usual way. Mr Stockmeyer gave me a small room in the Waldorf School with a table and a chair. Dr Maier brought in a microscope and a microtome. I gathered the necessary ingredients and the work began.[26]

Lilly Kolisko's laboratory experience from Vienna was helpful in this task. The correct extent of roasting coffee beans first had to be determined. She was already familiar with the production of microscopic specimens. Rudolf Steiner was at her side as a companion and helper:

> Today, as an independent person, I can vouch for the work that went out into the world from the Biological Institute at the Goetheanum. It was truly not an attempt to prove Dr Steiner's assertions quickly and at any price, but an honest striving for knowledge and truth. To achieve this, one will incur errors. Guided by Dr Steiner's wise and kind hand, it was possible to conduct research as a scientist in the anthroposophical sense with complete individual freedom. He paved the way everywhere: we just have to follow.[27]

Lilly and Eugen Kolisko carefully studied all the available literature on foot-and-mouth disease. She later formulated her attitude to the existing natural science in an article in the magazine "Natura":

> It is not possible to work anthroposophically in the right sense if one does not at the same time know exactly what science has to say about the subject in question. Dr Steiner never took up arms against the great achievements of modern natural science. He always and everywhere recognised them, but he could not stop where today's science has set its limits of knowledge; he led the natural sciences into spiritual science. For this reason, it was still possible to get answers from Dr Steiner to questions where modern natural science could no longer find answers.[28]

Lilly Kolisko tried all possible histological staining methods, but could not find anything conspicuous in the samples, and when Rudolf Steiner came to visit one day, she asked him to take a look at them. He described that in some samples individual cells were empty, while in others "little stars" were visible. But she couldn't see any sign of these.

However, she was always willing to learn new things: following Rudolf Steiner's instructions she travelled to Leipzig to learn microphotography from Professor Römer in order to be able to better study and identify the samples.

> After a few days I travelled to Leipzig to see Prof. Römer, who took me in warmly and showed and explained everything I needed. I had taken sections with

me to photograph them there. The excitement was great.

We increased the magnification from time to time, finally ending up with the oil immersion eyepiece 18 and the full extension of the apparatus. We used the highest possible magnification (approx. 3000) [...]. I stared at the plate - nothing, absolutely nothing could be seen. So all my efforts were in vain. Prof Römer found nothing either. I was none the wiser, and the photographer's camera had no better vision than I did. Now I stood in front of the window and turned the plate back and forth at some distance from the eye, more in play, I would say. The light fell on it in different ways – what was that shadow? In the centre of the protoplasm, a shadow in the shape of an oblique cross. That had to be the structure we were looking for. But now the joy was great. I was gleaming when I returned home from Prof Römer's. And now that I knew what the thing actually looked like, I also managed to find the change in the microscopic specimen. But it wasn't quite as easy as I had hoped and imagined.

You have to study it for a long time, train your eyes for a long time until you can see what really matters. [...] It took months of uninterrupted labour through long days and nights. Everything has to be conquered. [...]

I continued taking photographs back in Stuttgart. My little room, which consisted of a table, a chair, my microscope and a microtome, had grown in the meantime. I was lucky enough not to be put in a full room surrounded by lots of unnecessary complicated devices, but I was able to gradually acquire what I needed. Der Kommende Tag generously provided everything I needed. I owe it a great debt of gratitude because it trustingly made all the purchases that became necessary over the course of time. I did not remain alone but was initially supported by Miss Verena Gildemeister and later by Miss Kreuzhage [...].[29]

Germination experiments as a basis for the production of potentised remedies

The potency of the injected remedy was also investigated according to

18

Rudolf Steiner's instructions: "Carry out germination tests with the substance and represent the result in curve forms. Then the graph will give you a picture of the vitalisation process that the remedy exerts in the animal."[30]

This instruction signalled the beginning of ground-breaking research for anthroposophical medicine. Lilly Kolisko developed a method that demonstrated the impact of potentised substances. In the 1924 epilogue to her second publication on the smallest entities, she wrote:

> This publication is the result of four years of painstaking research. I would now like to reproduce, not literally, what Dr Rudolf Steiner said about four years ago, in March 1920, to the doctors and medical students[31] gathered around him in Dornach in response to a question about the effectiveness of homeopathic remedies:
>
> "With a homeopathic remedy, everything actually lies in the preparation. If, for example, you prepare silica up to high potencies, then you are working towards a certain point. In nature, everything is basically based on rhythmic processes. One works through a process towards a certain zero point. During this process, the initial effects of the substance come to light. Finally, one reaches the zero point where the effect of the ponderable substance is no longer manifest. If you then potentise further, it is not the case that everything disappears, but rather that the opposite emerges and is worked into the surrounding medium. So, if I were to say that a substance has certain properties, as I bring it to smaller and smaller quantities, then as I approach a known zero point, it acquires another property, that of radiating its former properties into the environment. If you potentise, you first reach a zero point. Beyond this point lie counter-effects.
>
> But that's not all; along the path that lies beyond this zero point, you can again arrive at a zero point - a zero point for these opposing effects. Then, by going beyond this point, you can arrive at even higher effects which, although they lie in the same direction as the first line, are of a completely different nature. *It would be a fine task to present the effects that emerge during potentiation graphically.*"[32]

Lilly Kolisko continued to write:

> What honest person striving for truth and realisation cannot be shaken by this fact? In these words, Dr Steiner gives nothing less than a description of the curves which I obtained after years of research. *It was indeed a marvellous task to depict in curves the effects that emerged from the potentiation of various substances.* Anyone who was able to see these words confirmed day after day could once again carry out their research work with genuine enthusiasm, could once again have the feeling when entering the laboratory: we are entering a consecrated place and the laboratory table becomes an altar. You feel that with every experiment you carry out you confront a riddle, and if you approach this riddle in a reverent mood, in devotion, then the whole cosmos responds to you through the experiment.

At the end of her book, she wrote: "*The curves are images of spiritual realities brought down to the physical. They reflect world laws.*"[33]

Lilly Kolisko worked intensively on this task. However, she had a long way to go before the curves became apparent.

> I had to keep searching for the protoplasmic structure and secondly, I had to carry out germination experiments. I approached the germination experiments with the same naivety. First I tried everything that is called germination experiments in normal scientific practice. The results were dismal. I took the highly concentrated preparation and put various seeds in it. Peas, lentils, beans, wheat. A well-meaning friend also advised me to use poppy seeds! But they were a bit too small for me. After a short time I gave up on the peas, lentils and beans and restricted myself to the wheat grains. In the concentrated solution, the grains withered and went mouldy after a few days. Then I diluted the preparation in an equal volume of water, then half more water again in the next glass dish ... and so on for perhaps 10 to 20 dishes. I hadn't even realised that there was such a thing as a [homeopathic] potency. I was delighted with the green plants, which were larger in one bowl and smaller in another. But then came the problem of visualising this growth in the form of a

curve. There was not much choice but to measure the individual leaves and the roots. The problem wasn't so bad with the leaves - there was only one or two; the plants couldn't grow any further in the small bowls. But there were plenty of roots. A whole clump! What should I do with the roots? Measure every single one? That seemed too much for me. Without further ado, I decided to take the longest root from each tuft. That's how my first curve was created. Dr Steiner came back ... I presented my curves and pointed to the growing plants in the glass bowls to see if this was what the task was meant to be. Dr Steiner replied: "Yes, that's what I meant, but you will probably have to dilute even more." Dr Steiner even agreed with this way of measuring, saying that I should only choose the longest root and not measure every single one. "You have to measure according to some principle, It's quite good like that. But you have to take it further and dilute it more."[34]

Sorting the wheat grains

A decisive factor for these trials was that the wheat grains should have the same inherent germination power. Ernst Lehrs described how Lilly Kolisko selected them:

> Rudolf Steiner first instructed her to try this with the means she knew. So she set about meticulously measuring the length, circumference and weight of the individual grains.
> It did not lead to any useful results. When she turned to Rudolf Steiner, he asked her to hand him a pile of grains and sorted them out with his hands to the right and left. (One was reminded of the fairy tale of Cinderella: the good ones in the pot, the bad ones in the bowl). The "good" ones all turned out to have the same germination capacity! Astonished, she asked him how he managed to achieve this. His answer: "With imagination." - "But I don't have that yet! So how can I do the work you want me to do?" - He then showed her everything that can be seen from thorough examination

of a grain of wheat: a dark spot, a fine feather, a certain lustre, etc. If she allowed herself to become more aware of this, she would soon realise which grains had same germination power. In fact, she largely achieved this![35]

In his book *Knowledge of Higher Worlds: How is it Achieved?* Rudolf Steiner developed the following exercise:

> Let the student place before himself the small seed of a plant, and while contemplating this insignificant object, form with intensity the right kind of thoughts, and through these thoughts develop certain feelings. In the first place let him clearly grasp what he really sees with his eyes. Let him describe to himself the shape, colour and all other qualities of the seed. Then let his mind dwell upon the following train of thought: "Out of the seed, if planted in the soil, a plant of complex structure will grow." Let him build up this plant in his imagination and reflect as follows: "What I am now picturing to myself in my imagination will later on be enticed from the seed by the forces of earth and light. If I had before me an artificial object which imitated the seed to such a deceptive degree that my eyes could not distinguish it from a real seed, no forces of earth or light could avail to produce from it a plant." If the student thoroughly grasps this thought so that it becomes an inward experience, he will also be able to form the following thought and couple it with the right feeling: "All that will ultimately grow out of the seed is now secretly enfolded within it as the force of the whole plant. In the artificial imitation of the seed there is no such force present. And yet both appear alike to my eyes. The real seed, therefore, contains something invisible which is not present in the imitation." It is on this invisible something that thought and feeling are to be concentrated. (Anyone objecting that a microscopical examination would reveal the difference between the real seed and the imitation would only show that he had failed to grasp the point. The intention is not to investigate the physical nature of the object, but to use it for the development of psycho-spiritual forces.)
>
> Let the student fully realize that this invisible something will transmute itself later on into a visible plant, which he will have before him in its shape and colour. Let him

22

ponder on the thought: "The invisible will become visible. If I could not think, then that which will only become visible later on could not already make its presence felt to me." Particular stress must be laid on the following point: what the student thinks he must also feel with intensity. In inner tranquillity, the thought mentioned above must become a conscious inner experience, to the exclusion of all other thoughts and disturbances. And sufficient time must be taken to allow the thought and the feeling which is coupled with it to bore themselves into the soul, as it were. If this be accomplished in the right way, then after a time—possibly not until after numerous attempts—an inner force will make itself felt. This force will create new powers of perception. The grain of seed will appear as if enveloped in a small luminous cloud. In a sensible-supersensible way, it will be felt as a kind of flame. The centre of this flame evokes the same feeling that one has when under the impression of the colour lilac, and the edges as when under the impression of a bluish tone. What was formerly invisible now becomes visible, for it is created by the power of the thoughts and feelings we have stirred to life within ourselves. The plant itself will not become visible until later, so that the physically invisible now reveals itself in a spiritually visible way.[36]

From then on, Lilly Kolisko was able to combine her spiritual training and her scientific work in practice. Later, Rudolf Steiner also gave a "seed meditation" for farmers. Adalbert von Keyserlingk wrote about this in his "Memories of early research work":

We worked with the exercise Rudolf Steiner gave which is to look at a seed grain lying in front of one and perceive the flame that may develop from it as one meditates. We would collect ears that seemed suitable for further development, take them to the laboratory, and then begin our meditation work. It was not a matter of looking at a single grain but of finding the few in a small pile of about a hundred seeds where we had the impression that their aura was particularly luminous. We would sow them in the ground to connect them with the powers of the earth and the powers of those among the dead who were prepared to help us. Our concentration and meditation made it possible for the dead and for spiritual entities to connect with the work we were doing.[37]

Determination of optimal shaking times for the potentiation of various substances

In the early 1920s Lilly Kolisko also addressed the question of the duration of shaking as a starting point for the potentiation of various substances. She developed a method for determining the optimum shaking time. She was already familiar with the capillary analytical method,[38] with which various substances are analysed for their composition and purity and modified it for this particular purpose.

In 1926 she wrote in the magazine "Natura":

> [...] For the first experiments we used the extract of the coffee bean, which had been prepared as Dr Steiner had indicated for the purpose of a remedy. Strips of filter paper, hung over 10 small glasses, were dipped into the coffee extract. Another 10 glasses contained the same extract, which had been gently shaken. (It was shaken up and down in a tube 28 times by hand.) It should show whether the shaking evoked a change in the rising height or in the shape of the edge of the pictures. After 14 hours, all the liquid had been absorbed and the paper strip had dried. There were indeed noticeable changes in the rise height and in the image. Many such experiments were now carried out, with ever increasing numbers of shakes (7, 14, 21, 28, 35, 42 etc.). During a visit by Dr Steiner to the Biological Institute, I was allowed to present the pictures. He examined them in detail and approved of the method. But the shaking was still insufficient. "You have to shake until you attain a horizontal upper line, then the material is homogeneous." In response to my naïve question about how long the shaking should last, Dr Steiner replied: "You'll just have to try it out." Now we had to work quite hard for weeks on end, first increasing the shaking very slowly by counting the shaking strokes. Attentive observation always revealed changes in the image between the individual shakes, but the horizontal line did not emerge. We shook all samples by hand. Finally,

24

we shook for longer durations of ½ minute, 1 minute, 1½ minutes, etc. up to 15 minutes. We ourselves were shaken all over. The arm muscles and head were particularly affected by this procedure. Our brains were always shaking in the end. But ... we reached the horizontal. When Dr Steiner returned, we were able to show him pictures that met with his approval. The material had become "horizontal" through the shaking.[39]

Lilly Kolisko described this method comprehensively in her publication Physiological and Physical Proof of the Efficacy of the Smallest Entities.[40]

Her work was of particular importance for the development of anthroposophical medicines. Hans Krüger confirmed this in a 1965 presentation at a Weleda pharmacist conference: "The optimal shaking times developed by L. Kolisko were immediately introduced into the practical production of medicinal products in the Weleda laboratories." [41]

Capillary Dynamolysis

In the early days of her collaboration with Rudolf Steiner in Stuttgart, Lilly Kolisko developed another important research method by extending the analytical capillary method.

In her book Agriculture of Tomorrow (1946), she described the emergence of what she called capillary dynamolysis (which other researchers later called the Rising Picture method):

At the Biological Institute at the Goetheanum (in Stuttgart), a specific method of investigation has been developed that makes it possible to discover the forces hidden in various substances. These "formative forces" are just as hidden in the materials as the "vitamins" are in fresh plant and fruit juices. We have named this method "capillary dynamolysis." It is based on the same principles as "capillary analysis" [...].

The scientist focuses only on the substances in capillary analysis [...]. We use the same principle, - liquids rise up filter paper - but we want to study the

25

forces hidden in the individual substances. *We want to look behind the veil of matter* [...]. This method, which we call *capillary dynamolysis*, has been developed over many years. In 1920, we began studying various metal salts; in 1923, we moved on to studying various plant saps, in accordance with a task set by Rudolf Steiner, whose philosophical works are known throughout the world, to the Biological Institute at the Goetheanum [...]. It took quite a long time before the solution to this seemingly simple problem—*to study the formative forces in plants*—with the help of filter paper and the corresponding plant extracts could finally be considered successful. Rudolf Steiner wrote a list of various plants whose formative forces were to be studied. We hope to one day be able to publish these experiments as well. They are not yet finished. Even after 20 years of work, we do not consider this particular work to be finished. *If we only use the plant saps, we will never discover the formative forces hidden in the plants.* After years of studying the formative forces hidden in inorganic matter, for example, in various metal salts, and their relationship to stellar forces, it was finally possible to find the solution to the problem: to reveal the formative forces in plants. *The effects of the metal salts must be linked to the various effects of the individual plants.*[42]

Rudolf Steiner was aware of the importance of this new method, capillary dynamolysis: Adalbert Keyserlingk, then a student at the Stuttgart Waldorf School, described it in his essay "Memories of Early Research":

> During a visit to our class, Rudolf Steiner told us that they had finally succeeded in grasping the etheric world in such a way that even materialistic science could no longer deny it, because these experiments could be repeated at any time. I have never seen Rudolf Steiner as joyful as he was there, when he was able to tell us about the success of the rising pictures.[43]

Elsewhere, Keyserlingk stated:

> In the early 1920s, I was able to learn about the work Frau Kolisko was doing in her Biological Research

Institute. This was in the school grounds a little distance away from the 'red wall' where the hall is today. We all knew and loved her, because she was so involved in and enthusiastic about her work. [...]. Frau Kolisko's biology research and Rudolf Steiner's lively interest in her work made a very deep impression on me in those days. We, the pupils of the upper school, were able to see a completely new method being tried out at the time, which was capillary dynamolysis. Countless series of experiments were done to create pictures in form and colour as fluids rose in filter paper and were then dried, making the ether forces visible for everyone to see. With breathless interest we found that it was possible to make the etheric nature of a plant visible by using its sap, or that of an animal by using its blood or lymph. We discussed the prodigious possibilities there would be in many fields, especially medicine, thanks to this method, and felt very much involved [...].

The most wonderful occasion was when Rudolf Steiner came into our classroom, his face alight with joy, and said: Now at last we are able to demonstrate the etheric, because of Frau Kolisko's work with very small entities, and can prove to anyone who wants to see that science can be taken further and can find its way out of the dead end of materialism!' I shall never forget the joy Rudolf Steiner radiated as he said these words. [...][44]

The influence of shaking on the substance Coffea praeparata

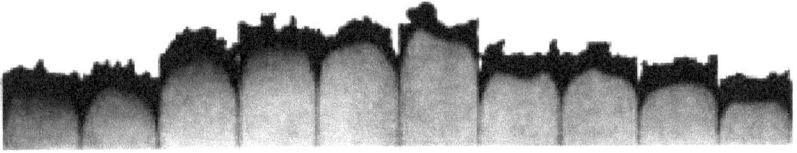

Unshaken

28 Shakes

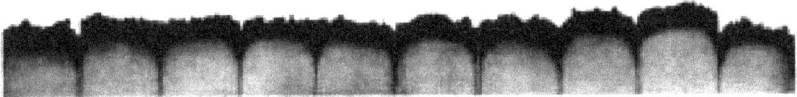

Shaken for 2.5 minutes

Fig. 8: Shaking time for coffee extract (Coffea praeparata). © Kolisko Archive, London, from Lilly Kolisko: Physiologischer und physikalischer Nachweis der Wirksamkeit kleinster Entitäten 1923 - 1959. Stuttgart 1959, p. 12

Function of the Spleen and Blood platelets

During the work on the therapeutic management of foot-and-mouth disease, another new area of research emerged. Lilly Kolisko began analysing blood samples from sick cows:

> Dr Kolisko went to the various stables affected by the epidemic to vaccinate the animals with the preparation. I often accompanied him, and as I always had a great interest in bacteriology (I used to study this speciality in Vienna at the General Hospital under Dr Bauer), I began to take blood samples from the animals to be injected, purely out of my own personal interest, and then again later after their injections. My thought was that perhaps we could use this opportunity to find the bacterium that causes this disease. I familiarised myself with all the relevant literature and collected many hundreds of blood smears which I subjected to various staining methods. Occasionally I told Dr Steiner about this and asked him to look at the microscopic preparations which I found extremely interesting. He politely but firmly declined each time. He believed me that they were interesting. Even when I told him that the red blood cells were completely destroyed, he only remarked that this was quite natural. During my private studies, I came across an element that was unknown to me and that I did not find mentioned in the numerous haematologies, with the exception of a picture in a large atlas. But in this picture of the blood of a cancer patient, it was categorised as a platelet. I didn't think that was right, and again I turned to Dr Steiner with the question: "I have found something in the blood of cows that I can't identify." Dr Steiner finally looked into the microscope and found it interesting. After brief consideration, he said that it was a hormone of the spleen. It could perhaps be called a "regulator". The same element would also be found in human blood if the rhythm of food intake were to be disturbed.
>
> Well, this statement by Rudolf Steiner interested me so I began my blood tests with people who volunteered to adhere to a dietary rhythm, or non-rhythm, prescribed

by me. These were mainly Waldorf teachers and the staff of the research institute "Der Kommende Tag", who were in the immediate vicinity so I could partly monitor the strict implementation of my instructions.

It was also possible to detect these "regulators" in human blood. This work ran parallel both to the investigations to find the structural changes in the protoplasm of the coffee bean that made the preparation a cure for foot-and-mouth disease in the first place, and to the plant growth experiments for determining the correct concentration of the remedy. Dr Steiner mentioned these investigations at the Clinical Therapeutic Institute, and this prompted the request that I should write about them.[45]

This work led to Lilly Kolisko's first scientific publication, *Function of the Spleen and Blood platelets*, in which she drew on the medical course[46] given by Rudolf Steiner in Dornach in April 1920 and the explanations it contained on the activity of the spleen:

"In this course, Dr Steiner explained, among other things, that the spleen is less connected with the actual physical metabolism than the other organs of the human abdomen, but that it is to a large connected with the *regulation* of the metabolism. It reacts to an extraordinarily high degree to the rhythm of human food intake and is ... to a high degree a *subconscious sensory organ* [...]. The activity of the spleen is therefore less orientated towards the actual metabolism in humans than towards the rhythmic processes."[47]

First university course in 1920 and second medical course in Dornach in 1921

The first anthroposophical university course in Dornach in 1920 (17 September to 16 October) was an important event for Lilly Kolisko. On Sunday, 16. September 1920, the so-called "opening ceremony" took place in the domed hall of the not yet fully completed First Goetheanum.

Lilly Kolisko described her experiences with the words:

For those who were lucky enough to attend the opening, it must have left an indelible impression. The prelude of the organ and orchestra roaring through the room, the solemn and expectant mood of the thousand or so people present, the coloured floods of light streaming in through the windows, the mighty columns, the wonderful paintings on the dome and Dr Steiner's deeply resounding words.[48]

During these weeks Rudolf Steiner gave a lecture course entitled "The Boundaries of Natural Science"[49], in which he described, among other things, a path of knowledge for natural scientists out of anthroposophy and the expansion of the capacity for knowledge into imagination, inspiration and intuition.

The topics of the course were wide-ranging and included philosophy, theology, history, linguistics, physics, mathematics, chemistry, medicine, education, art and economics. Among the many lecturers was Eugen Kolisko, who gave a lecture on "hypothesis-free chemistry" with a phenomenological orientation.

Lilly Kolisko, who later carried out important research work on these topics, was also present at the second doctors' course,[50] during which Rudolf Steiner spoke about questions of potentiation, medicinal plants and metal therapies, among other things.

Rudolf Steiner had the great expectation that the various areas of life and science would be deepened and renewed through anthroposophy.

From the summer semester of 1921 to the summer semester of 1923, Lilly Kolisko was enrolled at the University of Tübingen as a student of medicine and natural sciences[51], and from the winter semester of 1923 at the Technical University of Stuttgart (also natural sciences). Then, according to Margaret Bennell, she "had to give up her personal lifelong dream with much sadness and regret because of the ever-increasing tasks in the laboratory." [52]

Fig. 9: Postcard: The First Goetheanum, found in Lilly Kolisko's notebook.
© Rudolf Steiner Archive

Fig. 10: Group photo of the participants of the second medical course.
Lilly Kolisko in the bottom row, to the left of Rudolf Steiner, Eugen Kolisko to the
left behind Rudolf Steiner. April 1921 © Verlag am Goetheanum

4.

The publication of "Function of the Spleen and Blood platelets" for the Vienna West-East Congress (1922)

One always had to deal with currents and counter-currents that made productive work very difficult. It counted for nothing that some people researched and worked if others opposed it - including those working in the society. - because they did not understand it or believed they knew better ... even better than Rudolf Steiner [53]

An important milestone for anthroposophical work was the West-East Congress in Vienna from 1 to 12 June 1922. For Lilly Kolisko, this congress was the first touchstone as a researcher. She described her experiences as follows:

Dr Steiner had often pointed out that a *vademecum* was needed for medicine as well as for the other fields of knowledge. He repeatedly demanded this and was promised a medical *vademecum* for the West-East Congress. This promise could not be kept by the doctors. Instead, the Biological Department of the Scientific Research Institute Der Kommende Tag was called by the clinic and asked whether it would be possible to write a book about a work that Dr Steiner had spoken about at the Clinical Therapeutic Institute: "Spleen Function". For various reasons that are easy to understand, it was impossible to write a book at the last minute.

Shortly after the telephone call, Dr Steiner came to visit the research laboratory and was informed of this request. He seemed very tired, sat down and replied to my annoyed speech: "Oh yes, it would be very nice if you would." I replied that there was far too little time to write another book. I had never thought of writing anything about the investigations I had carried out in connection with the foot and mouth disease

34

investigations. There would hardly be time to print it either.

"Well, you'll just have to ring Der Kommende Tag, and they'll have to be in touch with the printers." The printers made it a condition that the manuscript had to be delivered ready for printing within two weeks at the latest and that no more author's corrections could be made. When I told Dr Steiner that I only had two weeks to look through all the material and write about it, he kindly said: "Well, I promise to read the manuscript immediately and send it back to you within 24 hours [...]." The book was written, approved by Dr Steiner and was printed for the West-East Congress [...].[54]

In her letter of 8 May 1922, Lilly Kolisko turned to Rudolf Steiner:

Honoured doctor!

The work on "Function of the Spleen and Blood platelets" has been compiled and is ready for printing as far as was possible in the short time available. I have negotiated with Mr Wachsmuth, Mr Palmer and the printer who hopes to have the brochure ready for the Congress. However, the manuscript should be delivered by Tuesday, 9 May. I very much hope to have the opportunity to discuss the work with you today, as I am not entirely sure that all of my remarks will meet with your approval. Unfortunately, I must go to Tübingen myself tonight and will return on Tuesday at 5 pm. I will leave the manuscript on your desk with the kind request that you review it.

Perhaps it would be possible for me to receive your notification sometime tomorrow, somewhere after 5 o'clock? If the work is to be on display at the Vienna Congress, then not another day must be missed. I apologise for the inconvenience. With devoted gratitude. Lilly Kolisko[55]

Fig. 11: Lilly Kolisko's letter to Rudolf Steiner dated 8 May 1922.
© Rudolf Steiner Archive

zurück. Das Manuskript lasse ich auf Ihren
Schreibtisch deponieren, mit der Bitte um gütige
Durchsicht.

Vielleicht liesse es sich ermöglichen, dass ich
morgen irgendwann, irgendwo nach 5ʰ um
Ihren Bescheid holen könnte? Soll die Arbeit auf
dem Wiener Kongress ausliegen, dann darf kein Tag
mehr versäumt werden. Verzeihen Sie gütigst die
Belästigung.

In dankbarer Ergebenheit

Lilly Kolisko

Lilly Kolisko recorded what happened next with the words:

> I had no intention of travelling to Vienna myself. Very shortly before the Congress, Dr Steiner gave another lecture for members in Stuttgart. He approached me in the anteroom, greeted me and said: "I've heard you're going to give a lecture in Vienna. That will be good." Well, I hadn't heard anything about it and thought it was a mistake, a confusion of persons, so I declined. "Yes, it will," replied Dr Steiner. "That will be so good." The following day I telephoned the Clinical Therapeutic Institute to find out what Dr Steiner's statement could refer to and was assured that it was true, that the doctors had suggested that since I had written the book, I could also give a lecture on it. It was the first book I wrote and the first lecture I presented. Dr Steiner had approved the manuscript, he was in agreement with the lecture, and immediately afterwards the doctors at the Clinical Therapy Institute congratulated me on the book. It still rings in my ears when Dr Palmer called out to me at a meeting: "This is a book that has a head and a foot "das Kopf und Fuß hat". I read it in one go. It's excellent."[56]

Function of the Spleen and Blood platelets was the first anthroposophical medical publication with a scientific-experimental orientation; it was also noticed by the public and in some cases sharply criticised. An article in the "Deutsche Medizinische Wochenschrift" ended with the words: "If *the only positive aspect of this whole brochure* is to assert the appearance of sharply contoured, dark platelets, apparently young forms, due to disruption of the rhythm of food intake, the author would do well to repeat the investigations on an exact basis and not with the equipment of Der Kommende Tag."[57]

Not all anthroposophical doctors found the work "excellent" or understood the new paths of anthroposophical research. This was clearly expressed during the "Medical Week" in Stuttgart in October 1922 (23 - 28 October).

»Der Kommende Tag«

Wissenschaftliches Forschungsinstitut

MITTEILUNGEN

HEFT 1

(Herausgegeben von der Biologischen Abteilung)

Milzfunktion und Plättchenfrage

von

L. Kolisko

1922

DER KOMMENDE TAG A.=G. VERLAG / STUTTGART

Fig. 12: Function of the Spleen and Blood platelets

39

In October 1922, a "Medical Week" took place in Stuttgart, organised by the Clinical-Therapeutic Institute. Dr Walter Johannes Stein had received permission from Dr Steiner to attend his lectures. I met him on my way to the lectures and he asked me why my book on "Function of the Spleen" was not on the book table. He said it was an opportunity for the book to be known among doctors. I didn't know and advised him to ask the member who had the book table. Perhaps she had overlooked it. After a few minutes, he returned excitedly with the information: "Frau Kinkel has been instructed by the Clinical Therapeutic Institute not to distribute the book. She should only sell it if it was specifically requested." I immediately went to Dr Steiner and told him what had happened. "If the book is bad, then it should be cancelled. If it's good, it shouldn't be taken off the book table."[58]

Rudolf Steiner wrote on 25. November 1922 from Stuttgart to Marie Steiner:

[...] I have work to do here from morning till night. And now, as I am writing this, Husemann, who is the instigator of the boycott of Kolisko's brochure, phoned me. I had to tell him over the phone because he wanted to speak to me in person today: An even more blatant opposition emanated from the College of Physicians than from anyone in the Society: You have given the order to bury a brochure that I think is good during Doctors' Week so that no-one can see it: I will not accept any explanations from you.[59]

Lilly Kolisko talks about her situation at the time and subsequent developments:

Rudolf Steiner came back to Stuttgart and spoke to me: "I've heard that you've been so depressed recently. Why is that?" "Well," I said "the matter of the spleen book is very close to my heart. Either the book is bad, as I've already said, in which case it should be pulped. Or the book is good, in which case it shouldn't be opposed."
"Come and see me at 3:30 pm in Landhausstraße. We must talk about it."

After much consideration I have decided to publish something about these details, because it must be of value to the members of the anthroposophical movement to look into the difficulties that existed within the Society in 1922 and which gradually led to Rudolf Steiner refounding the Society in 1923 - taking over the leadership of the Society himself. You always had to deal with currents and counter-currents that made productive work very difficult. It counted for nothing that some people researched and worked if others opposed it - including those working in the society. - because they did not understand it or believed they knew better, even better than Rudolf Steiner. I look back on a conversation I had with one of the doctors after the book had been excluded from the doctors' meeting. "Don't you realise," I said at the time, "that you are not taking action against me at all, but actually against Dr Steiner? I haven't done anything special. I did find this element in the blood, but I could never have said that it is a hormone of the spleen, that is a regulator, that can also be found in human blood. Only Dr Steiner could explain that to me. So you are taking action against Dr Steiner when you attack this book." I received the following reply from the doctor, a prominent member of the medical profession: "Well, you know, Dr Steiner is not a haematologist." ... What can you do? The doctors were prepared to accept Dr Steiner's wonderful lectures on medicine, they had confidence in his knowledge of human nature, they brought him seriously ill people to ask for his advice, but they did not trust him to know as much or even more about human blood than a young doctor who fancied he knew something about haematology. That was actually in the background of my depression at the time. Perhaps this attitude of the doctors could also be included in Dr Steiner's frequently used complaint of "inner opposition".

Now I came to Dr Steiner in Landhausstraße and was asked to come into his room. I saw the doctors from the Clinical Therapeutic Institute sitting in the anteroom. Dr Steiner was very serious and asked me: "What do you yourself have to say about the book?" He wanted to hear my own judgement on it.

Now I said: "I've read the book over and over again, and if I had to write it again today, I couldn't change the content in the slightest. But, Doctor, you know that I had very little time to write this book, I could perhaps formulate it better stylistically."
"Then that's fine," he replied. "Then I'll demand that the doctors publicly recant." Now I felt sorry for the doctors, and I started to say: "But Doctor, you can't force the doctors to do that. They simply don't understand."
"They can say that they don't understand it. But they are not allowed to say that it is not scientific."
Once again, I tried to make an interjection: "But Doctor, I feel sorry for Dr ..." "That's your personal business, it doesn't belong here." That was the end of my conversation, which left me with a profound impression of the seriousness and relentless rigour with which Dr Steiner proceeded [...]. After some time an article about the book appeared in one of the anthroposophical journals, written by one of the doctors, who spoke favourably.[60] However, it was not written by the personality who had behaved particularly negatively.[61]

Rudolf Steiner spoke about Lilly Kolisko's "epoch-making" spleen studies in a lecture in Dornach on 22 October 1922 shortly before his departure for the "Medical Week" in Stuttgart, but expressed already his regret about how little this scientific work was recognised in the Anthroposophical Society:

Now, take an organ like the spleen. Ordinary physiology and medicine don't have much to say about it. You will find in all corresponding textbooks the notation: about the spleen one does not yet have anything to say today. You will find that everywhere, if you look it up. That is not very surprising. You see, the genius of language is really wiser in this respect than science. In this case - in other cases it is the German language genius which is extraordinarily wise - it is the English language genius which designates the (Milz) as "spleen". And that is an extraordinarily favourable designation, because the spleen is connected with all those activities of man which go beyond the ego, which approach the spirit-self. The spleen is even directly the organ of the spirit-self. It enters fully into the spiritual realm. Only one must be able to stand it. Most people cannot tolerate the real spiritual element. Therefore, they are not in any way animated through the activity of the spleen to an activity that is spiritual but become "spleeny". In reverse, they are tuned down. The "spleen" is nothing other than a spirit which, instead of going into the head, twists itself into the bowels. Therefore "spleen" is an extraordinarily good designation, which points directly towards the spirit, for which the spleen is the corresponding organ.

The spleen is effective in bringing about a balance, as presented in the pamphlet, - which has been worked out in our Physiology Institute, particularly by Frau Dr. K., - where the activity of the spleen is presented in relation to platelet formation and the whole digestive process.

This is truly scientific and systematic presentation of spleen activity in the first instalment. If such work were to be done in another research institute somewhere, it would very soon be presented as something extraordinarily epoch-making. But the fact is that when something is created in our circle, in the bosom of our society, it doesn't get out into the world. People don't talk about it. It's not necessary to talk about it in order to praise it, but because it could have a beneficial effect in the context of the whole of contemporary society. But the beginning of not talking about it is already being made in our Anthroposophical Society. I would like to take a vote on how many of our members have had the opportunity to really realise the full significance of the matter! It is not surprising then that when the Anthroposophical Society begins to pay no attention to what is happening within us, this naturally has an effect on the outside world. In fact, we are not only working to the exclusion of the public, but also to the exclusion of the Anthroposophical Society's interest in the most important things! But that is what I want to say - today at least - only in parenthesis."[62]

43

In 1922 Rudolf Steiner wrestled intensively with the leaders of the Anthroposophical Society about the future, and the question of science played a major role in this.[63] However, he did not achieve a turnaround. Steiner's disputes, including with the medical council of the Stuttgart Clinical Therapeutic Institute (Friedrich Husemann, Otto Palmer, Felix Peipers, Ludwig Noll), were recorded as follows in a meeting of the Stuttgart "Circle of Thirty", four weeks after the fire at the Goetheanum on 1 January 1923:

> What good is advice if things go on like this? When you let the most inappropriate happen unmonitored at the most important moments? What good is advice if I have been saying for months that I would like to hear why this happened, that the spleen brochure was boycotted? What good is advice? I must not hear what prompted the College to order that no eye must see the brochure! I must not hear why these things are like this! It doesn't help to talk about giving advice. That is part of what is ruining society. How different our scientific endeavours would be today if one of the doctors had opened his mouth and said something that has been sought for God knows how long! You can publish ten remedies with insubstantial claims! But if the world had learnt that these things had been done in a clinic, the whole world would have been talking about it. Why doesn't something like this happen? Why is something like this not being talked about, even though I have been asking for it for weeks? Why is it being kept quiet? They will follow all my advice and boycott it. Why is that? The Anthroposophical Society has developed to such an extent that one could say that there is internal opposition: Inner opposition is being made; for example, among those to whom it would have fallen to promote the spleen brochure. The Anthroposophical Society has allowed a circle to come into open opposition to me. And this despite the fact that I have repeatedly made it clear that everything I have said has been ignored. Is it acceptable that a medical course is being held here, and then what immediately appears to be a significant achievement, is being boycotted? Does one appreciate the scandalous nature of this matter? This makes one say: The Society is doing nothing. The question is: Will the Society now engage

in such a way that I will no longer be slapped in the face by the Anthroposophical Society as before?

Dr Rascher is staying in Dornach in a house where Frau Häfliger lives and she learns a lot from him about the opposition to the spleen brochure. I ask you: How am I, how is such a matter treated even in narrow circles? How did the medical profession feel responsible for the very thing that it has committed itself to keeping within its circles? That is the Anthroposophical Society! - It must have happened very quickly. Think of this embarrassment. I'm always being molested that I should give permission for people to read the doctors' courses.

Dr Rascher: I would like to ask the doctors if they don't want to answer.

Dr Husemann: It was because we feared what was in the brochure. I was afraid of the discussion. It happened out of cowardice.

Dr Steiner: If we continue to do things this way ... [gap in shorthand] I have not yet seen a review of Frau Dr Kolisko's brochure in "Anthroposophie". The path you have taken is to make it disappear, perhaps revive it in a clinic ten years later. Study the history of German scholarship in the 19th Century to see what happened there. I really haven't held back with positive advice recently. None of it has been followed. It is a matter of giving advice at one point and then having it all ignored. [64]

On 4 January 1923, a few days after the fire at the Goetheanum, Steiner also said in Dornach:

The opposition is not dormant. It can only be countered with positive achievements of the society. Because scientists have emerged in recent years the society has to start with what it wants to continue in the outside world. But if we do it in such a way that we are so poorly disposed to our own work, we will never consolidate the society. It is necessary to bring about conditions in the society itself so that the achievements support each other. The conditions around the Kolisko brochure are to the ruin of the society. [65]

Despite all this, Eugen and Lilly Kolisko continued their intensive work in Stuttgart, including their research into foot-and-mouth disease in the Biological Department of "Der Kommende Tag", using capillary dynamolysis among other things. Eugen Kolisko also used this method for medical-diagnostic purposes as a school doctor at the Waldorf School, treating over 1200 children over the years. The experiments on the "smallest entities" were continued with comprehensive potency series. During several visits to the Biological Institute, Rudolf Steiner provided new ideas for future work, including further spleen investigations. For example, the spleen of hares or rabbits were to be removed by surgery, as can be read in a calendar entry by Eugen Kolisko: "Spleen from hares = Rabbit surgery - 'ether spleen'".[66] Lilly Kolisko began these animal experiments, which were later mentioned by Rudolf Steiner in several lectures.[67]

She wrote in retrospect: "Despite the hostility and attacks, the investigations into spleen function were continued, and a year after the publication of the first work, a second was completed. However, Rudolf Steiner refrained from publishing it after the first one had been so badly received, even by those who should have welcomed it with joy."[68]

5.

The Christmas Conference 1922/23
and the Goetheanum fire

"Despite this I would like you to give another lecture on the spleen during the Christmas conference, followed by a discussion. I want to give the doctors another opportunity to talk about it.[69] I had previously stood upstairs and looked amongst the smoking columns. Dr Steiner was standing next to me and I asked:"Doctor, do I have to give the lecture?" He turned his head towards me and said: "I'll speak too."[70]

On 31st August 1922, Lilly Kolisko turned 33 years old. The Christmas conference in 1922-23 and the dramatic events of the night of the fire were deeply shattering experiences for her. There are two different descriptions of the events by Lilly Kolisko, the first from 10 October 1928 reads as follows:

It was New Year's Eve 1922, and the last words after Rudolf Steiner's lecture still ring like a warning in our ears:

> *"In Earth-activity draws near to me,*
> *Given to me in substance-imaged form,*
> *The heavenly Being of the Stars—*
> *In Willing I see them transformed with love!*
>
> *In Watery life stream into me,*
> *Forming me through with power of substance-force,*
> *The heavenly Deeds of the Stars—*
> *In Feeling I see them transformed with Wisdom."* [71]

When we left the Goetheanum it was with the brightest hopes for the future. Just in the afternoon of the day before, a resolution was passed to perform the Mystery Plays again.

In this New Year's Eve atmosphere, while the weather was cold but still refreshing, people wandered home in groups, chatting happily. People had come from all over for this Christmas conference, so Dornach was full to bursting. - - -

A group of people were walking in a very cheerful mood towards the Clinical Therapeutic Institute, where the Christmas tree was to be lit among friends. Under the Christmas tree, dinner was ready. They went to the table and the mood was the same, alternating with serious moments, but still upbeat and very happy. Suddenly a telephone call rang out, which sounded very long, shrill and almost alarming (In those days, the ringing signal was generated by turning a crank on the caller's telephone set. The caller could not only change the length of the tone, but also the volume by turning the crank quickly).

And we were not at all surprised when a minute later a nurse rushed into the room, pale as death, and called out: ... the phone, the Goetheanum is on fire, and we should bring in all the Minimax equipment to put it out. Suddenly an indescribable shock, dead silence, but then a rush out of the house; everyone armed themselves with a Minimax or any kind of instrument that might possibly help. The street was already swarming with people, as the cry "the Goetheanum is on fire" had spread everywhere.

They raced up the hill. And when the Goetheanum appeared for the first time and you saw a cloud of smoke hanging over it, this moment could probably be described as if something had been inflicted on the dearest thing you had. You had the feeling that something had happened to a loved one. If possible you quickened your steps in order to help, even though there seemed little hope, because everyone there knew that the Goetheanum is a wooden building. With clouds of smoke already rising from the building, the danger was great. When we got to the hill, there were already a lot of people working there and the fire brigades had been called from all around, so our Minimaxes were no longer an option. We went into the building with the Minimax. Despite our best efforts, we couldn't get the fire under control. And we had to watch - all those gathered around the building and Rudolf Steiner

48

between us - as the flames burst out of the dome at exactly twelve o'clock midnight. Half an hour later the roof collapsed with a great roar, and now
we could see that that the Carpentry workshop was also in danger from the sea of flames. The Carpentry workshop where work was still being done and Dr Steiner's studio and the bookstore and the eurythmy cloakroom were all located. All attention now focused on the Carpentry workshop. The books were taken away to a neighbouring house. They also began to clear out Dr Steiner's studio. Here was the statue of Christ and of Lucifer and Ahriman. It was only possible to move the statue of Christ out by breaking through an outside wall and then carrying the superhuman-sized statue through this opening.

The top part of the statue was carried out in three parts. It was more and more painful because you could see that everything was lost. It was desolating when the fire brigade gave up the struggle to save the building because they considered it hopeless. Only then did they douse the Carpentry workshop with torrents of water, as the heat was so great that they feared that the Carpentry workshop would also ignite.

It was an incredible spectacle to see a large building collapse piece by piece. The glass windows melted under the heat and you could see the marvellous flames. The organ melted.

The flames were metallic red and green and blue. And then you could see the tall pillars, twelve in number, burning like torches up into the sky. An open mystery site was consumed in fire. Silently, deeply moved, the members stood around and waited until the last pillar fell.

On all sides was a jeering crowd of people, who had not yet reached the site, whizzing cars of curious people who had rushed up from everywhere to watch the spectacle. People stayed together until fatigue and the pain drove many back home. But Rudolf Steiner did not leave. He kept wandering around the grounds, retreating to the nearest hut for half an hour, only to reappear when there were only very, very few people left on the site. And so, I was lucky enough to meet Rudolf Steiner in the morning hours around 5:30. We stood almost all alone on the grounds and looked at the sad remnant of the Goetheanum, which had been so proud just a few hours ago.

"It is deeply sad what we have had to experience," Rudolf Steiner said to me. I took his hand because I couldn't reply as my voice failed me. "We must not be discouraged and continue tomorrow exactly as we had planned." [72]

Lilly Kolisko's second description of the night of the fire can be found in her book about Eugen Kolisko, which was published in 1961. It says there:

The Christmas Conference came with Dr Steiner's wonderful lectures: "The Origins of Natural Science". The members' lecture on 24th December on the revelation of the secrets of the course of the year, on 30 December the lecture on the position of the movement for religious renewal in relation to the anthroposophical movement and on 31 December, 1923 the last lecture held at the Goetheanum, "Spiritual Knowledge Is a True Communion.".

Fig. 13: Ruins of the first Goetheanum. Dornach, 1 January 1923.
© Hoffmann Photo Kino AG, Basel

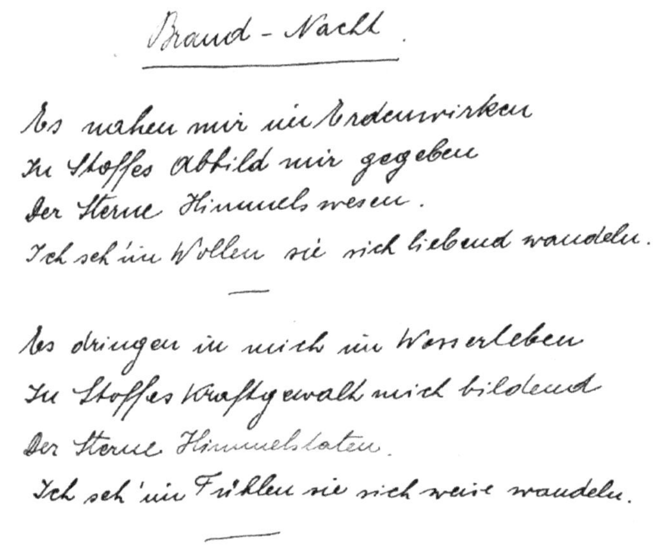

Brand - Nacht

Es nahen mir die Erdenwirken
In Stoffes Abbild mir gegeben
Der Sterne Himmelswesen.
Ich seh' im Wollen sie sich liebend wandeln.

Es dringen in mich die Wesenleben
In Stoffes Kraftgewalt mich bildend
Der Sterne Himmelstaten.
Ich seh' im Fühlen sie sich weise wandeln.

Fig. 14: Burning night, Rudolf Steiner. From Lilly Kolisko's meditation book.
© Rudolf Steiner Archive

Shortly afterwards, the Goetheanum was in flames. Many reports were written about the fire. Dr Kolisko and I were in the canteen when the first shouts arose: The Goetheanum is on fire! We ran up the hill and couldn't see anything yet. One of the members came out of the entrance, said something about smoke in the White Hall and collapsed. More and more members had rushed up, a chain was formed and buckets of water were passed from hand to hand. They still didn't know where the source of the fire was. Dr Steiner supervised the efforts and gave various instructions. He instructed me to prepare a table for first aid in case it was necessary. Dr Kolisko was nowhere to be seen. The fire brigade took a long time to finally intervene to help. Friends came to me from time to time and said: "We can't find Dr Kolisko anywhere, Dr Steiner is very worried about him. We know that he went into the Goetheanum, but nobody saw him come out. Those members of the Anthroposophical Society who spent the night of the fire on Dornach Hill on New Year's Eve 1923 will probably never forget it. They had to give up hope when the flames burst out of the domes, the coloured windows melted and then the pillars lit up like torches burning up to the sky. Finally, the news came that Dr Kolisko was safe. He later told us that he followed the traces of smoke in the Goetheanum and, standing in the large auditorium, saw the flames engulf the curtain. He wanted to stop the flames from spreading by tearing down the curtain and ran up the gallery. Suddenly the lights went out and in the growing smoke and darkness he had great difficulty finding his way back. It took a long time before he came back outside. He was the last one in the burning Goetheanum.

The work continued the next day. At 5 o'clock in the afternoon, Dr Steiner spoke before the start of the Three Kings Play in the Carpentry workshop. It was a tremendous imposition on the actors. My task was to speak to the assembled doctors in the Glass House about the function of the spleen, followed by a discussion. I had previously stood upstairs and looked into the smoking rubble. Dr Steiner was standing next to me and I asked:

"Doctor, do I have to give the lecture?" He turned his head towards me and said: "I'll speak too." I don't remember what I said at the time. I also don't know whether the doctors present were able to concentrate on the lecture. Their eyes kept wandering out to the smouldering pillars. There was also a discussion, I wrote down Dr Steiner's answers and when I read them again today, I find them wonderful. Dr Steiner's attitude probably helped us all to uphold our duty.[73]

6.

The second publication:

The Physiological and Physical Proof of the Efficacy of the Smallest Entities

If you hold to this method, which has grown out of anthroposophy itself, then you will not need to lose heart. [74]

After the Goetheanum fire Lilly Kolisko continued with her various research tasks including writing a second publication on potentiation and plant growth experiments. During a visit to Dornach before the Christmas Conference she had consulted Rudolf Steiner about a title: "He thought a little and then said: 'The physiological and physical proof of the efficacy of high dilutions'. Then after a short pause: 'No ... better: 'The physiological and physical proof of the efficacy of the smallest entities.' An entity is something much more comprehensive than a high dilution" [75]

Later, in 1959, she wrote about her collaboration with Rudolf Steiner in a major publication:

In 1923, I was in the fortunate position of being able to present the manuscript to Dr Steiner for review. He had provided the impetus for this research and had followed the progress of the work with interest. He listened patiently when I came up with my objections, when I was not satisfied with the results, indeed, he spent many hours in the laboratory, and finally he carried out an experiment with me, supervising the preparation of potencies, the selection of the seeds, the cleaning work that took place in between, the method of shaking, in short, observing every phase of the experiment to see if there was a possible source of error somewhere. Finally, I was able to convince myself that the experiments were flawless, and only then did Dr Steiner say: *"Well, let's publish them now."* After he had read through the manuscript, he sent it back with a friendly letter saying that it was good and could be printed.[76]

The publication was issued in April 1923.[77] Three months earlier, in January 1923, shortly after the fire at the Goetheanum, Rudolf Steiner had said in Stuttgart in a critical discussion with the anthroposophical doctors and social leaders:

> [...] Here in Stuttgart there has never been the will to work in a phenomenological way, except in the Biological Research Institute, where two series of experiments have come out that hold up. If you stick to this method, which has grown out of anthroposophy itself, then you will not need to lose courage. Bringing in the university methods does not work. It really is a matter of taking responsibility for what is in harmony with anthroposophy. It is a question of how to make fruitful progress and not of endless series of experiments that lead nowhere [...] [78]

Lilly Kolisko's work attracted attention in various homeopathic journals. An article appeared in "Die Allgemeine Homöopathische Zeitung" in December 1923 in which her methodology - from the selection of the wheat seeds to the preparation of the soil for the samples - was discussed extremely critically: "[...] We cannot but say that we have never in recent times found published experiments in which all the rules of exact scientific research have been disregarded as much as here [...]." [79] A book review in the journal "Heilkunst", on the other hand, came to a positive assessment:

> [...] The absolute reliability of these results can no longer be doubted, given the care with which all sources of error were excluded in the experiments. The curves included in this work provide a more vivid picture of the individual test results than the description can. [...] Homeopathy must be particularly grateful to the author for providing, in a scientifically unassailable form, definitive proof that the potencies of her remedies can no longer be simply treated as ineffective "nothings." The solutions used are partly beyond the "Avogadro constant", which undoubtedly shows that physiological efficacy is by no means tied to the presence of chemically detectable amounts of substances. [80]

In conversations with Rudolf Steiner, new questions arose in Lilly Kolisko's mind and she repeatedly received suggestions for her further practical work. On 30 November 1923 she wrote a long letter to Rudolf Steiner with detailed descriptions of her latest experiments:

> Highly esteemed Doctor!
> In the following, I would like to take the liberty of giving a brief report on my latest attempts, as you are unlikely to be in Stuttgart before Christmas.
> When you were last here in October, you set me the primary task of obtaining curves where the weight curve shows the opposite tendency to the growth curve when exposed to intensive light. The aim was to show that light counteracts gravity. It gives me immense pleasure to be able to inform you that a good part of this task appears to have been solved [...].
> [...] The experiment with gold chloride, which you saw in October, also produced very nice curves.
> I continued the capillary analysis experiments and tried to achieve coloured mercury pictures by combining iodine potassium and mercury nitrate or mercury chloride.
> Perhaps it will be possible, when I come to Dornach at Christmas, to present the curves on gold and the capillary analysis pictures to Herr doctor? I would be very grateful if I could receive your kind advice for the continuation of the work [...]. [81]

This work was the basis for her next publication on the "smallest entities", which was to appear three years later in 1926. [82]

7.

The Christmas Conference 1923/24

A scientific impulse will have to emanate from the Anthroposophical Society. This must be made evident at the moment when we want to take the Anthroposophical Society into entirely new channels [...].[83]

We have here laid the Foundation Stone. On this Foundation Stone shall be erected the building whose individual stones will be the work achieved in all our groups by the individuals outside in the wide world.[84]

The impulse of the Christmas Conference - with the laying of the spiritual foundation stone of the Anthroposophical Society here in this hall – means it was indeed the case, as I said yesterday, that an esoteric current would have to flow through the whole Anthroposophical Society from now on, an esoteric current that could already be noticed in everything that is been attempted within the Anthroposophical Society since Christmas. The core of this esoteric work of the Anthroposophical Society must now be the esoteric school [...].[85]

The Christmas Conference of 1923/24[86] brought about major changes in Lilly Kolisko's life. After it, the Biological Department of the Scientific Research Institute of Der Kommende Tag was affiliated both ideally and really with the Goetheanum, the School of Spiritual Science in Dornach.

Lilly Kolisko became a member of the First Class of the School of Spiritual Science - and for the faculty of the Waldorf School in Stuttgart she facilitated the esoteric Class lessons held by Rudolf Steiner. She later took on this task for an even larger group of members of the School of Spiritual Science in Stuttgart.[87]

For the penultimate day of the Christmas Conference, 31 December 1923, she had been asked to give a presentation on her scientific work.[88] Rudolf Steiner had cited her work as an example of what would be necessary to make progress in anthroposophical-scientific research.

On this day, 31 December 1923, Rudolf Steiner first read the entire Foundation Stone Meditation and then the rhythm belonging to this day, which was written on the blackboard:

Light divine,
Christ-Sun
The spirits of the elements hear it
from
East, West, North, South:
May human beings hear it![89]

After these words, physicist Rudolf Maier, head of the Scientific Research Institute of Der Kommende Tag AG, gave a presentation entitled "The connection of magnetism with light".

Rudolf Steiner then spoke about the importance of scientific work within the Anthroposophical Society and said, among other things:

My dear friends, it will be of the greatest importance that a truly anthroposophical method should be made customary in the different branches of scientific life by those individuals who are called to these branches within our anthroposophical circles. Indeed, seen from a certain point of view, this is of the utmost importance. If you seek the source of the great resistance of our time that has been appearing for decades against any kind of spiritual-scientific view, you will find that this resistance comes from the different branches of natural science. These different branches of natural science have developed in isolation, without any view of the world in general.

Round about the middle of the nineteenth century a general despair began to gain ground in connection with an overall view of the world. People said: All earlier overall views of the world contradict one another, and none of them has led anywhere; now it is time to develop the sciences purely on an exact foundation, without reference to any view of the world.

Half a century and more has passed since then, and now any inclination to unite a view of the world with science has disappeared from human minds. Even when scientific research itself urges an attempt to be made, it turns out to be quite impossible because there is insufficient depth in the spiritual-scientific realm.

If it should become possible for Anthroposophy to give to the different branches of science impulses of method which lead to certain research results, then one of the main obstacles to spiritual research existing in the world will have been removed. That is why it is so important for work of the right kind to be undertaken in the proper anthroposophical sense.

Today there is an abyss between art and science; but within science, too, there is an abyss between, for instance, physiology and physics. All these abysses will be bridged if scientific work is done in the right way in our circles. Therefore from a general anthroposophical point of view we must interest ourselves in these different things as much as our knowledge and capacities will allow. A scientific impulse will have to emanate from the Anthroposophical Society. This must be made evident at the moment when we want to take the Anthroposophical Society into entirely new channels. [...].[90]

After Rudolf Steiner's speech, Lilly Kolisko gave her presentation and spoke about the scientific work of the Biological Institute in Stuttgart: "Proof of the efficacy of the smallest entities". Rudolf Steiner then took the floor again and emphasised his great appreciation of Lilly Kolisko's research work and its importance for the future:

Well, my dear friends, you have seen that quiet work is going on amongst us on scientific questions and that it is indeed possible to provide out of Anthroposophy a stimulus for science in a way that is truly needed today. But in the present situation of the Anthroposophical Movement such things are really only possible because there are people like Frau Dr Kolisko who take on the work in such a devoted and selfless way. If you think about it, you will come to realize what a tremendous amount of work is involved in ascertaining all these sequences of data which can then be amalgamated to form the curve in the graph which is the needed result.

These experiments are, from an anthroposophical point of view, details leading to a totality which is needed by science today more urgently than can be said. Yet if we continue to work as we have been doing at present in our research institute, then perhaps in fifty, or maybe seventy-five, years we shall come to the result that we need, which is that innumerable details go to make up a whole. This whole will then have a bearing not only on the life of knowledge but also on the whole of practical life as well.

People have no idea today how deeply all these things can affect practical daily life in such realms as the production of what human beings need in order to live or the development of methods of healing and so on.

Now you might say that the progress of mankind has always gone forward at a slow pace and that there is not likely to be any difference in this field. However, with civilization in its present brittle and easily destructible state, it could very well happen that in fifty or seventy-five years' time the chance will have been missed for achieving what so urgently needs to be achieved. In the face of the speed at which we are working and having to work, because we can only work if there are such devoted colleagues as Frau Dr Kolisko—a speed which might lead to results in fifty, or perhaps seventy-five years—in the face of this speed, let me therefore express not a wish, not even a possibility, but merely, perhaps, an illusion, which is that it would be possible to achieve the necessary results in five or ten years. And I am convinced that if it were possible for us to create the necessary equipment and the necessary institutes and to have the necessary colleagues, as many as possible to work out of this spirit, then we could succeed in achieving in five or ten years what will now take us fifty or seventy-five years. The only thing we would need for this work would be 50 to 75 million Francs. Then we would probably be able to do the work in a tenth of the time. As I said, I am not expressing this as a wish nor even as a possibility, but merely as an illusion, though a very realistic illusion. If we had 75 million Francs we could achieve what has to be achieved. This is something that we should at least think about.[91]

The Biological Institute at the Goetheanum

In 1926, Lilly Kolisko explained how important changes affecting her research institute in Stuttgart came about after the Christmas conference:

> At first, the room I worked in was called the "epidemic department". This name was later changed to the "Biological Department", and then after the Christmas Conference, when Dr Steiner founded the General Anthroposophical Society, he changed the name to "Biological Institute at the Goetheanum". It was a memorable conversation that we [Dr Kolisko and I] had with Dr Steiner and Dr Wegman in the studio in Dornach. Dr Steiner said: "So you want to be completely connected to the Goetheanum in Dornach? Well, I have always regarded you as belonging to the Goetheanum." Dr Wegman then suggested moving the Biological Institute to Dornach. Dr Steiner said that unfortunately that would not be possible because of the Waldorf School, but that laboratories would also be built into the new Goetheanum, and then in future it would probably have to be that you would carry out the preparatory work here in Dornach together with me and then do the work in Stuttgart. "The Biological Institute has developed in a healthy way. It has grown out of the work. It wasn't that the rooms were full first and then you looked for the people for them, but the work was there and you started to work, and then it grew out of the work. The Biological Institute was healthy from the start and will continue to thrive [...]." [92]

The contract for the changes was only signed on 6 March 1924, after weeks of discussions in Stuttgart, including with the representatives of Der Kommende Tag and the Clinical Therapeutic Institute. Lilly and Eugen Kolisko reported on the negotiations in a letter to Ita Wegman dated 16 January 1924:

> Esteemed Frau Dr Wegman!
> Herewith we take the liberty of reporting on the nego-
> tiations held in Stuttgart with Mr Leinhas regarding the

the incorporation of the Biological Institute at the Goetheanum. Dr Kolisko sought to speak with Mr Leinhas immediately after his arrival in Stuttgart on 7 January, but was only able to meet on Thursday 10 January. In this conversation, Dr Kolisko essentially put forward the points of view discussed in Dornach regarding the transfer of the Biological Institute from Der Kommende Tag. Mr Leinhas did not seem to take the matter very seriously and was of the opinion that the whole matter could be clarified in a discussion between him and Dr Steiner [...].[93]

Lilly and Eugen Kolisko also wrote in the letter about another meeting with Leinhas. This conversation shows that Rudolf Steiner's comments about the Biological Institute during the Christmas Conference had left their mark:

> [...] It was also clear that he did not really believe that the desire for a complete - including financial - separation came from Dr Steiner himself. Among other things, Mr Leinhas was of the opinion that if Dr Steiner pulled out the department of the research institute that was just about to prove useful and only wanted to leave what was a burden for him to Der Kommende Tag, Der Kommende Tag would not be in a position to manage the other departments. Mr Leinhas said that Dr Steiner was certainly satisfied with the Biological Department and wanted to take it over, but not with the working methods of the other research staff at the Research Institute. In this case, Der Kommende Tag could have no interest in continuing to finance these institutions [...].

Further discussions had been held, including a meeting with the board of the German Anthroposophical Society. They wanted to wait in Stuttgart before making a decision and to clarify the matter further until Rudolf Steiner's next visit to Stuttgart. In the end, however, Lilly and Eugen Kolisko wrote to Wegman: "In any case, we remain fully committed to the proposals agreed in Dornach and are endeavouring to implement them [...]."

The following contract was finally signed on 6 March 1924:

> Acquisition of the Biological Department of the Scientific Research Institute from "Der Kommende Tag AG", Stuttgart, by the Goetheanum, Dornach.

Agreement between the company "Der Kommende Tag Aktiengesellschaft zur Förderung wirtschaftlicher und geistiger Werte, Stuttgart", represented by its Board of Directors, and the Goetheanum, the School of Spiritual Science in Dornach, represented by the Executive Council of the General Anthroposophical Society.

1. the Biological Department of the Scientific Research Institute "Der Kommende Tag" in Stuttgart will be taken over by the Goetheanum, the School of Spiritual Science in Dornach, with effect from 1 January 1924 and will henceforth be renamed the "Biological Institute at the Goetheanum". The previous employees of the Institute, Herr Dr Kolisko, Frau Dr Kolisko and Sister Ruth Kreuzhage, will now enter the service of the Goetheanum.

2. The Goetheanum will reimburse Der Kommende Tag from the German Fund for the Reconstruction of the Goetheanum one third of the amount (calculated in gold marks) that has been spent by Der Kommende Tag on the establishment and operation of the Biological Department since its foundation until 31 December 1923. The stocks of goods and the inventory of the Biological Department thus become the property of the Goetheanum.

3. The operation of the Biological Department from 1 January 1924 will be charged to the German Fund for the Reconstruction of the Goetheanum. The payment of salaries and expenses will be settled by Der Kommende Tag, which in turn will be charged to the German Fund for the Reconstruction of the Goetheanum. The rooms used in the Scientific Research Institute of Der Kommende Tag will continue to be made available to the "Biological Institute at the Goetheanum"; the rent for them will be calculated according to the same principles according to which the rent is charged to the other departments of the Scientific Research Institute on the basis of the statutory rent regulations applicable in Germany.

Stuttgart, Dornach, 6 March 1924
For the Goetheanum, the School of Spiritual Science:
The Executive Council of the General
Anthroposophical Society
Dr I. Wegman, Secretary, Rudolf Steiner
Chairman Der Kommende Tag, Aktiengesellschaft,
Leinhas [94]

Fig. 15: View from the driveway, on the left the teachers' house, on the right the Koliskos' Research Institute. From: Dietrich Esterl: "Die erste Waldorfschule Stuttgart Uhlandshöhe", edition waldorf

Fig. 16: Lower school garden with Lilly Kolisko's Biological Institute on the right and the Delmonte, Ruthenberg and Michels houses in the background. From: Dietrich Esterl: "Die erste Waldorfschule Stuttgart Uhlandshöhe", edition waldorf

Biologisches Institut am Goetheanum
der
Freien Hochschule für Geisteswissenschaft in Dornach

Stuttgart, den 31. Okt. 1927.
Kanonenweg 44ᴵᴵ

Fig. 17: New letterhead. © Rudolf Steiner Archive

Lilly Kolisko's new stationery with the letterhead "Biological Institute at the Goetheanum of the School of Spiritual Science in Dornach" bears witness to the affiliation with the Goetheanum, which also secured the finances in the years to come.

The School of Spiritual Science

The esoteric school announced by Rudolf Steiner at the Christmas Conference in 1923/24 as the basis for all future work at the School of Spiritual Science[95] became essential for Lilly Kolisko. As early as 6 January 1924 she sent her application for admission to the School of Spiritual Science to Rudolf Steiner. Her letter was formulated very simply and without "presentation"; Steiner had known and appreciated her since their first meeting in Vienna in 1915.

> Dornach, 6 January 1924, Honoured Dr Steiner, please accept my application for admission to the School of Spiritual Science.
> In grateful respect,
> Lilly Kolisko [96]

Lilly Kolisko attended the first lesson of the First Class of the School of Spiritual Science on February 15, 1924, and from then on travelled weekly to Dornach for each class lesson, initially held on Friday evenings and later on Saturdays—and the associated member lectures. During her stays in Dornach, she regularly had the opportunity to speak with Rudolf Steiner about her research; she also forwarded questions and reports from Stuttgart about the Waldorf School and anthroposophical work to him. She also attended the Class lessons in Breslau during the Agricultural Conference in Koberwitz. She took the work on the mantras for the Class lessons very seriously and strove to "be a worthy representative of the anthroposophical cause before the whole world with all her thinking, feeling, and will." [97]

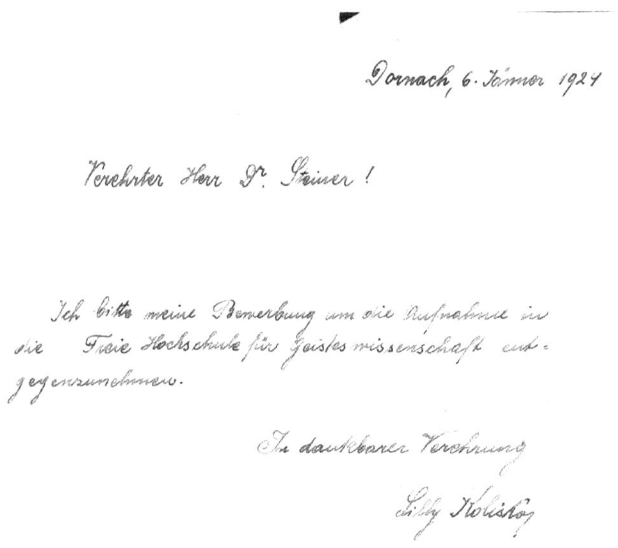

Dornach, 6. Jänner 1924

Verehrter Herr Dr. Steiner!

Ich bitte meine Bewerbung um die Aufnahme in
die Freie Hochschule für Geisteswissenschaft ent-
gegenzunehmen.

In dankbarer Verehrung

Lilly Kolisko

Fig. 18: Lilly Kolisko: Application for admission to the School of Spiritual Science. 6
January 1924 © Rudolf Steiner Archive

Lilly Kolisko already realised the importance of this work after the first class lesson - and the question arose within her: How could more people learn about this content? She dealt with this inner question and later wrote about her considerations:

> The newsletter mentioned several times by Rudolf Steiner in the course of 1923 appeared on 13 January 1924 and was entitled "What is happening in the Anthroposophical Society". The individual members were to participate in a spiritually lively way in everything that was going on in the Society, but also what was happening in contemporary spiritual life outside the Society was to be included in the common consciousness. Dr Steiner began to write about the "School of Spiritual Science", the first lecture of which took place in Dornach on 15 February 1924 for those members who had received their blue certificates by this time and were present in Dornach. Anyone who was lucky enough to hear Dr Steiner himself on these occasions will know what an unprecedented experience these lessons given to the 1st Class were for the participants. Since I myself was entrusted with a specific task within the First Class, and since various, partly completely untrue, representations have been given about it over time, which continue to circulate within the Society to this day, I would like to include the true representation here. This is an important area within the Anthroposophical Society.
>
> After attending the lesson, I was extremely impressed and asked myself the question: What about all the other members who do not live in Dornach and therefore do not hear the magnificent messages given by Dr Steiner? Could there be a way of making the content of these lessons accessible to other members? I turned to the Secretary of the Society, Dr Wegman, with my question. "Yes, how do you envisage that?" she replied, and I replied that perhaps at least a small circle in Stuttgart could be told about it, perhaps a circle like the "Dreißiger Kreis". Dr Wegman promised to ask Dr Steiner about it, and then Dr Steiner called me in. I want to reproduce this conversation as faithfully as possible. Dr Steiner said: "Dr Wegman informed me of your intention to convey the Class lessons to a circle of members. I like it

very much. But why does it have to be the Dreißiger Kreis? The Dreißiger Kreis is not an institution with which I can work esoterically." I replied that the Dreißiger Kreis wasn't particularly important to me, I had only named this institution to express my wish to be able to convey the content to at least a small group of people. Dr Steiner then continued: "Wouldn't you like to pass it on to the Waldorf school teaching staff?" Of course I would be happy to do so, and Dr Steiner promised to issue the necessary membership cards for the entire teaching staff immediately. Thus began this organisation for the Stuttgart teaching staff. Dr Steiner asked me if I wanted to do it, and I happily accepted his proposal.

This assignment meant that I travelled to Dornach each week to attend the lectures for members and Class lessons. Naturally, I had many questions about how to hold the lessons. I didn't feel able to convey the content of the lessons in my own words and asked Dr Steiner if he would allow me to take notes. He gave me permission, although no one else was allowed to take notes. I expressly say "no one". I will come back to this later. I was able to talk to Dr Steiner in the studio almost every week about matters concerning the First Class, about my research work, about matters concerning the Waldorf School, which often had questions forwarded through me, and about general Society matters. [98]

The facilitation of the Class lessons in Stuttgart expanded after a while to a larger circle of members. She reported the following:

Sometime after I had shared the class content at the Waldorf School, I had a conversation with Dr Steiner in the studio in Dornach. He told me that other endeavours were now also underway to hold the Class lessons for all members in Stuttgart. Mr Arenson and Dr Unger had contacted him. I then asked whether the two gentlemen wished to do this. "No," replied Dr Steiner. "They don't even want it for themselves. They've chosen someone else to do it." (Again, I emphasise that I am reporting this conversation verbatim) "And that is ..." I don't want to give the name. In any case, I was astonished and asked if the person in question could do that.", it would be equivalent," came Dr Steiner's reply. "I told the gentlemen that if it is at all possible to communicate it to all members of the class in Stuttgart, then only *one* person would be considered, and that would be you. My only concern is that it might be too much for you to hold two lessons a week in Stuttgart. But people don't want you, so there's nothing we can do ..." Again, I wondered why Dr Steiner had told me this. Some time passed without me hearing anything more about it. Then suddenly Dr Unger approached me at a meeting in Dornach, a folder in his hand and said: "Yes, right, what I wanted to tell you is that the list of members for the general Class Lesson is much larger than I had originally assumed. I was just on my way to the studio, so the first thing I said to Dr Steiner was: "Dr Unger has just told me that the list of members of the First Class in Stuttgart is much larger than he had thought. I wonder if that means that the gentlemen have agreed that the lessons will be transmitted by me? Dr Steiner: "So, has he done that? Well, then I will arrange it now ..."

On 10 April 1924, a meeting of representatives was held in Stuttgart, at which Dr Steiner referred to this:

"[...] Then there are also other institutions. For example, Frau Dr Kolisko always gives the Class lessons for the teaching staff of the Freie Waldorfschule, including a few other friends from Stuttgart. Another group of people who strive for this is being formed here in Stuttgart. Isn't that right, Mr Arenson (Mr Arenson stands up and says 'Yes'). And if such groups are formed in the near future, Dornach will do what is necessary to find the appropriate communications and facilitation. I don't think it's really appropriate for people to declare that they are fit for such a service. Because I don't believe that anyone can judge for themselves exactly whether they are appropriate for such a service. But I do believe that those who are unsuitable consider themselves suitable. That cannot be taken amiss, it is also quite understandable [...]." [99]

In a letter to Rudolf Steiner dated 22 June 1924, Lilly Kolisko wrote about the importance of the connection between the Waldorf School and the spiritual sources of the Goetheanum. She also told him about the organisation of Class lessons in Stuttgart:

> Esteemed Dr Steiner!
> In the following I would like to take the liberty of giving a brief report on the meeting of the members of the first class of the School of Spiritual Science (insofar as they belong to the teaching staff of the Waldorf School in Stuttgart), which had become necessary after the last conference we were able to have with you.
> All colleagues were present with these exceptions: Dr Schwebsch, Miss Tillis, Miss Gildmeister, Mr Baumann.
> Mr Baumann did not want to come because he had the feeling that the Doctor had not yet decided his case and it would therefore be better if he kept away from all class matters until a decision had been made.
> The aim of the meeting was to find ways and means of making such unpleasant events impossible.
> In a few introductory words I pointed out once again the significance of the fact that everyone had registered twice, as an individual and as a teacher of the Waldorf School. With this act the Waldorf School had placed itself under the esoteric leadership of the School of Spiritual Science in Dornach. I pointed out the significance of the fact that from now on the Executive Council members of the General Anthroposophical Society present in Stuttgart would take part in the teachers' conferences, in short, how a completely different position of the Waldorf School in relation to the anthroposophical movement had arisen. Every individual must bring this fact to vivid consciousness.
> If Dr. Steiner, in his kindness, has now given permission for a continuous stream to flow to Stuttgart
> from the spiritual spring bubbling up in Dornach, then we have a responsibility to ensure that this happens in a dignified manner.
> I would just like to briefly outline the results of the discussions. It was decided:
> 1. the lectures now take place without exception on

Thursdays at 9 o'clock in the evening in the singing hall of the Freie Waldorfschule. Should any event make it impossible to give a lecture on a Thursday, the lecture for that week will be cancelled and held on the following Thursday.

2. Mr Stockmeyer is responsible for notifying the individual class members, both inside and outside the school, and will ask for the help of some colleagues if necessary.

Mr Wolffhügel and Mr Strauss will take over the room watch and will receive a list of all participants for this purpose. Mr Wolffhügel will close the hall at 9 a.m. sharp. Late-comers will not be admitted.

Mr Stockmeyer shall arrange for the collection of the travel amounts in such a way that he opens an account "for anthroposophical purposes", which is held at my disposal. The individual colleagues will have the corresponding small amount deducted directly from their salary each month. For the other participants, we will try to ask Dr Palmer to set up a similar facility for the Clinical Therapeutic Institute and then transfer the total amount to the account "for anthroposophical purposes"; the same applies to the participants of the Christian Community, where Dr Rittelmeyer will probably take over, then a few individuals remain who can transfer it directly.

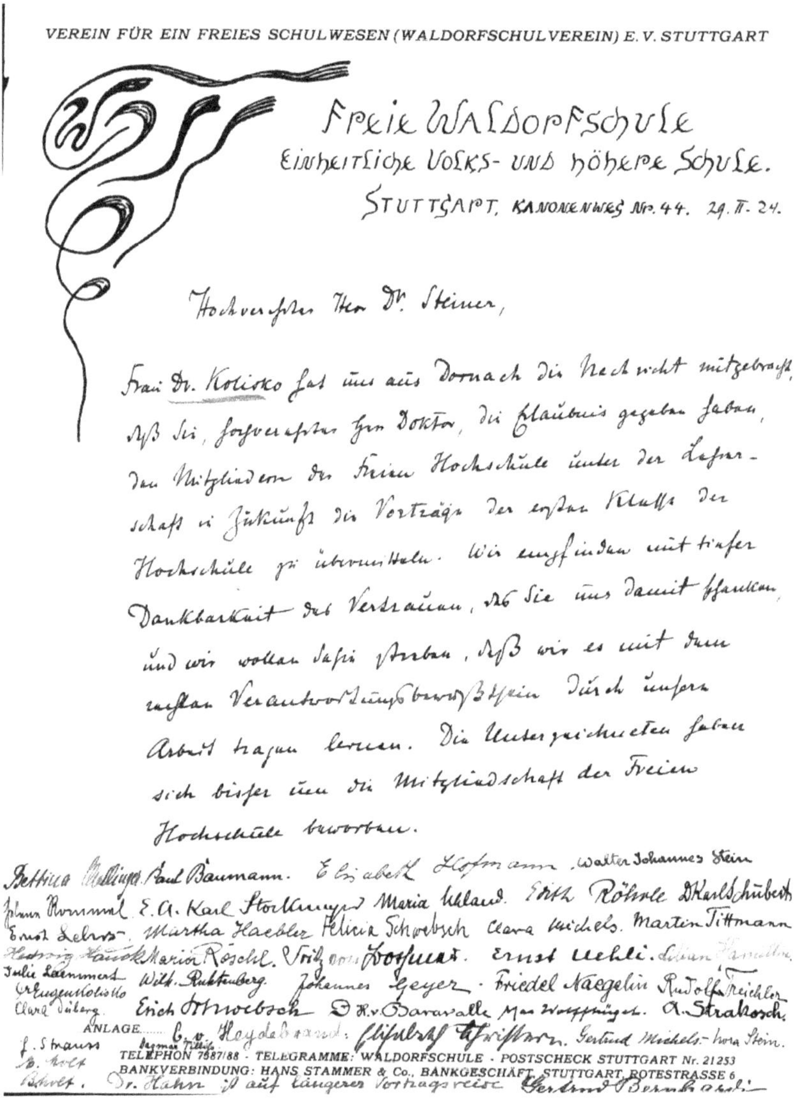

VEREIN FÜR EIN FREIES SCHULWESEN (WALDORFSCHULVEREIN) E. V. STUTTGART

Freie Waldorfschule
Einheitliche Volks- und höhere Schule.
Stuttgart, Kanonenweg Nr. 44. 29. II. 24.

Hochverehrter Herr Dr. Steiner,

Frau Dr. Kolisko hat uns aus Dornach die Nachricht mitgebracht, daß Sie, hochverehrter Herr Doktor, die Erlaubnis gegeben haben, den Mitgliedern der Freien Hochschule unter der Lehrerschaft in Zukunft die Vorträge der ersten Klasse der Hochschule zu übermitteln. Wir empfinden mit tiefer Dankbarkeit das Vertrauen, das Sie uns damit schenken, und wir wollen dafür sorgen, daß wir es mit dem rechten Verantwortungsbewußtsein durch unsere Arbeit tragen lernen. Die Unterzeichneten haben sich bisher um die Mitgliedschaft der Freien Hochschule beworben.

Bettina Mellinger. Paul Baumann. Elisabeth Hofmann. Walter Johannes Stein
Johann Rommel. E. A. Karl Stockmeyer. Maria Uhland. Edith Röhrle. Dr. Karl Schubert
Ernst Lehrs. Martha Haebler. Felicia Schwebsch. Clara Michels. Martin Tittmann
Hedwig Hauck. Marion Röschl. Fritz von Bothmer. Ernst Uehli. Lilian Hamilton
Julie Laemmert. Wilh. Ruhtenberg. Johannes Geyer. Friedel Naegelin. Rudolf Treichler
Dr. Eugen Kolisko. Clara Düberg. Erich Schwebsch. Dr. H. v. Baravalle. Max Wolffhügel. A. Strakosch.
ANLAGE ...... E. v. Heydebrand. Elisabeth Schwietzer[?]. Gertrud Michels. Nora Stein.
J. Strauss. Dagmar Tillich. B. Wolf. Baravalle[?]. Dr. Hahn ist auf längerer Vortragsreise. Gertrud Bernhardi.

TELEPHON 7587/88 · TELEGRAMME: WALDORFSCHULE · POSTSCHECK STUTTGART Nr. 21253
BANKVERBINDUNG: HANS STAMMER & Co., BANKGESCHÄFT, STUTTGART, ROTESTRASSE 6

Fig. 19: Letter of thanks from the members of the teaching staff of the Stuttgart Waldorf School to Rudolf Steiner dated 29 February 1924.
© Rudolf Steiner Archive

It is now necessary to communicate this arrangement to the other participants in the Class presentations. It has been decided to invite those participants who are not members of the teaching staff to a meeting before the next presentation, Wednesday, June 25th, at which the entire teaching staff will be present. This is necessary because the various participants have gradually arrived and may have some confusion regarding the establishment of the entire program, particularly regarding travel expenses. Since Dr. Steiner has now declared the list of participants closed, this can be easily carried out.

At the end, I spoke a few words about the tremendous responsibility the teaching staff would have to bear if they wanted to be the vessel into which the spiritual wealth flowing from Dornach would flow, and that Dr. Steiner now also expected something from the teaching staff. Something should now also flow from the teaching staff into the sad atmosphere of Stuttgart. I believed that it was the task of the teaching staff to destroy the "Stuttgart system," which had caused so much harm, and that we should gradually succeed in allowing something to radiate into the atmosphere of Stuttgart that would enable Dr. Steiner to hold Class lessons here himself.

Unfortunately, I now have to add something to the report. I had read out the list of participants to the staff, and now some of them came forward to say that Frau Dr Husemann had been present in the previous Class Lesson. She was not noticed in the chaotic atmosphere. As she is not on the list and the list was marked as closed by Herr Doctor, she will be informed accordingly.

I will also inform Mr Leinhas that he will not be able to attend for the time being.

We would now be grateful to Dr Steiner if we could hear whether the measures, we now want to take meet with his approval.

With regard to the matter of Mr Baumann, I would ask to hear what is necessary before Thursday, if possible. With grateful respect L. Kolisko [100]

The last of the 19 Class lessons was given by Rudolf Steiner in Dornach on 2 August 1924. After a trip to England [101], he gave seven more Class lessons, the so-called "recapitulation lessons" between 6 and 20 September in Dornach, which were attended by many new class members. Rudolf Steiner personally admitted almost 250 people to the School in the weeks of September.

On 17 October 1924, during the first period of Rudolf Steiner's sick leave in his studio, Lilly Kolisko wrote to Ita Wegman about the further work with the Class lessons in Stuttgart:

> [...] At school, I've reached the point of the first recapitulation lesson (6 September, 1924) with the Class presentations. May I return to my last conversation with you and ask what the plans are for the presentations for all class members in Stuttgart? Now that the repetition is beginning, one could well bring about a reunion. There is a great longing for this in many hearts. Of course, I don't want to prejudge Dr. Steiner's decisions with an immodest request and leave it entirely up to your discretion whether or not to discuss this matter with Dr. Steiner. I would also like to thank you from the bottom of my heart for the helpful indications about my treatment. In gratitude and respect, Yours, Lilly Kolisko.[102]

Lilly Kolisko's letter of 2 November 1924 to Ita Wegman shows that from then on, she began to read the Class lessons for all class members in Stuttgart. She also wrote about her observations of the effect of the Class lessons on the teaching staff:

> Esteemed Frau Dr Wegman!
> I have received your kind letter and thank you very much for it. It is a tremendous gift that is given to the entire Stuttgart membership. Of course, I cannot conceal from myself the great responsibility it places on us, but we must take this responsibility upon ourselves with courage.
> When I look at the teaching staff before Christmas and after Christmas, for example, it is undeniable that a lot has changed, which is solely due to the fact that the Class lessons have engendered an intimate bond around the whole staff. The faculty has really become something

in recent times. I would like to say that there is a sense of seriousness, of holy seriousness, which is necessary for such a sacred thing. "The faculty should form a nucleus," Herr Doctor told me at the time, "from which something can radiate." I believe that this core is becoming stronger and stronger. You will also know that a meeting of anthroposophical teachers takes place almost every Sunday after the Services for the children. There, too, something moves gently through the room that can fill you with satisfaction. Many questions can now be discussed that could not have been discussed before Christmas. Something is gradually emerging from the chaos that is full of life.

We can therefore proceed with courage and hope to give presentations to the other class members as well. I believe that the whole of Stuttgart will breathe a sigh of relief, as if relieved of a heavy nightmare, when it hears that this permission could be given.

But there are still details to be discussed. I will consult with some friends about what can be done for card control, hall security, etc. Furthermore, it is still unclear what will happen to those members who have not yet received their blue card. Some very old members are said to be among them. Can these cards be issued, or is this completely impossible at present? Should an invitation to attend also go out to those members who live in the immediate vicinity of Stuttgart, e.g. in Esslingen, Ludwigsburg, Pforzheim etc., or should only those from Stuttgart come?

Furthermore, what happens if class members from outside the area happen to come to Stuttgart? Does possession of the blue certificate alone entitle one to admission, or is participation limited to local members?

These are all questions that still remain to be answered and clarified. As soon as I have discussed the security measures with my friends, I would like to submit them to you for review. I am coming to Dornach this week and would like to start the week after next, Thursday 13 November, in Landhausstraße, if you agree [...].[103]

Rudolf Steiner and Ita Wegman agreed. Clarita Berger wrote to Ita Wegman's colleague Mien Viehoff and the clinic friends in Arlesheim four weeks later, on 10 December 1924:

> [...] Do you know that Frau Dr Kolisko repeats the Class lessons here in the branch? She does it very well, reads the shorthand very vividly. Does everything like the Doctor - and is carried by a holy seriousness [...].[104]

The "young doctors"

A third area in which Lilly Kolisko was very actively involved after the Christmas Conference was medicine, particularly in the work of the so-called "young doctors" - a group of medical students and young doctors centred around Helene von Grunelius and Madeleine van Deventer – a group to whom her husband Eugen Kolisko would later also belong. [105]

On 5 December 1923, Helene von Grunelius wrote to Madeleine von Deventer:

> "Dear Maddy, [...] I am enclosing the final (i.e. from our side quite final) list [...]. Kurt Magerstadt is Hardt's friend. I invited Frau Kolisko as a guest after I had a long, very fine conversation with her last Sunday. However, I realise that we have no reason to invite her husband. After all, there is a big difference between the two of them. If Dr Steiner wants to invite him, that is something else and none of our business [...]." [106]

The first "Jungmediziner" course, which began immediately after the end of the Christmas Conference, was actually the first course of the new School of Spiritual Science and had a mantric sequence; it showed how Rudolf Steiner wanted to begin the work of the School in the specialist sections anew and deeply impressed the listeners.[107] After the course, Lilly Kolisko reported to Ita Wegman in a long letter dated 15 February 1924 about the work of a small group in Stuttgart and their studies:

> Dear Dr Wegman, [...] You will perhaps be most interested in what the participants of the medical course

in Stuttgart are doing, and I would like to give you as brief an account as possible. Before this course was presented, a group of young people had worked here and studied the lectures on "Occult Physiology". One day these young people then asked me to take part in the meetings, and these were then once a week in the Biological Institute. The participants were: Helene v. Grunelius, Dr Bort, Dr Altenmüller, Miss Hachez, Ammerschläger and Rudolf (two students from the Technical University), Miss Leschke (a teacher who was both an intern at the Waldorf School and painted together with Frau Strakosch) and Mr Linde (who had a few semesters at the Agricultural University). After the Christmas course in Dornach, the question arose as to how these meetings should be continued. On the one hand, we wanted to process the suggestions received in Dornach; on the other hand, it was not really possible to convey the content of the lectures to those young friends who were not doctors and had not taken part in the course. Either the course would have had to be split up or it would have fallen apart altogether. It now seemed a pity to exclude Ammerschläger, Rudolf, Linde and Leschke, as they really did participate in the meetings with a lot of good will and diligence. I now made the following suggestion: at Christmas, Dr Steiner once again pointed out with great emphasis that we should first and foremost seek knowledge thorough exoteric schooling before embarking on an esoteric deepening of the studies, otherwise we would just have to stumble along. Now I must admit that the exoteric knowledge of all participants is somewhat weak. As soon as they come into contact with anthroposophy, young people very easily tend to neglect the ordinary sciences and then fail to create a solid foundation for themselves. We will now try to provide ourselves with such a foundation, which is particularly necessary for non-medical people, and once we have this, only then can we proceed to study occult physiology. In this way, all participants would be served. The suggestion was applauded.

We have now set ourselves the first task of acquiring a vivid image of the four-fold human being: the solidly contoured, the fluid, the airy and the warmth-being. Dr Steiner said in his lectures that it is easiest for people today to understand the warmth being. So, it would actually have been a given to start with the warmth being and end with the solidly contoured. Now I must confess that I didn't manage to do that and started with the bones. Right and wrong, probably more wrong than right. On this occasion I came to a lecture in which Dr Steiner explains that it doesn't really make much sense to just look at the skeleton, you actually have to be able to hear the bones, you have to face the skeleton with a musical feeling. Of course, we are not yet capable of this and had to limit ourselves to standing before this marvellous structure with a sense of awe and admiration.

I began the first evening by making a more general observation about the whole solidly contoured human being [...] [108]

Lilly Kolisko also reported on the four other meetings and concluded the letter with the words: "[...] Furthermore, that we use the Saturday evening, if possible, to discuss purely medical issues in a smaller circle, following the Christmas lectures [...]." [109]

Her medical group also responded to the "Appeal to Youth" written by Rudolf Steiner, which appeared in the weekly journal "Das Goetheanum" on 24 February 1924 in the newsletter of the weekly magazine "Das Goetheanum":

ANNOUNCEMENT OF A YOUTH SECTION
Newsletter, 24 February 1924

The Executive Council of the Anthroposophical Society at the Goetheanum is seeking to establish not only the sections already mentioned but also a further section. This will be possible if the intentions of the Executive Council meet with a corresponding response. In every age, young people have been somewhat at odds with the old. This adage is a consolation to many when it comes to the behaviour of today's youth. But this consolation could easily become a disaster. One should understand the present youth from the "spirit of the present" both in their questionable aberrations and in their all too justified striving for something different from what the old give them. First of all, there is the youth that is pushed into the academic career by the circumstances of life. They are offered "science". Solid, secure, fruitful science for the outer life. It would be nonsense, in the manner of many laymen, to rant about this science. But the soul of youth still freezes to this science before it comes to recognize its solidity, its security, its fertility for the outer life. Science owes its greatness to the strong opposition it has faced since the mid-19th century. At that time, people realized how easily man can sail into the uncertainty of knowledge when he rises from the lowlands of research to the heights of a world view. It was believed that chilling examples of such a rise had been experienced. And so they wanted to free "science" from the world view. It should stick to the "facts" in the valleys of nature and avoid the high roads of the mind. When they opposed the worldview, they derived a certain satisfaction from the act of opposing. The worldview fighters of the mid-19th century were happy in their fighting mood. Today's youth can no longer share this happiness. They can no longer stir up satisfying feelings in their souls by experiencing the fight against the "uncertainty" and "crush" of the worldview. For today there is simply nothing left to fight against. It is impossible to advocate freeing "science" from "worldview." For the worldview is dead by now. In contrast, however, the feelings of young people have made a discovery. Not at all a discovery of the intellect, but one that comes from the whole, undivided human nature. The young have discovered that without a worldview, it is impossible to live a dignified human life. Many of the old have heard the "evidence" against the worldview. They have submitted to the power of the evidence. The youth no longer pay any intellectual attention to this power of evidence; but it instinctively senses the powerlessness of all intellectual proof where the human heart speaks from an invincible urge. Science presents itself to young people in a dignified way; but it owes its dignity to the lack of a world view. Young people long for a world view. But science needs young people. At the Goetheanum, we would like to understand young people in such a way that we can seek the paths to a worldview with them. And we hope that in the light of the worldview, a true love for science will be generated. We would like to not lose science in world view reverie, but to gain it in the awakening of spiritual experience.

The leadership of the Anthroposophical Society asks young people if they want to understand it too. If they find this understanding, then the "Section for the Spiritual Strivings of Youth" can become something vital. [110]

Lilly Kolisko then sent the following letter to Ita Wegman on 11 March 1924:

Honourable Frau Dr Wegman!

Please allow me to enclose a carbon copy of a letter to Dr Steiner, which I am enclosing. It is intended to express precisely what the young physicians would like to say about the appeal to young people. I now ask you to be kind enough to read the copy to see whether it corresponds to what Dr. Steiner expects. If you do not find the letter good and correct, then please do not forward it and let me know by Friday. A copy of the letter should be sent simultaneously to the other course participants, with the request that they also express their views in the same vein if they agree with its content. Yours sincerely, Lilly Kolisko. [111]

The enclosed letter to Rudolf Steiner said:

Stuttgart, 11 March 1924,

Honourable Herr Dr Steiner!

The Stuttgart participants of the course that took place at Christmas in Dornach for younger doctors and physicians feel compelled to say something heartfelt in response to the appeal to young people that was made in Newsletter No. 7. Every single word in the appeal to young people spoke to our hearts. Yes, the knowledge of science is dignified, but cold and sober. It leaves young people dull. Today you can no longer live life as a scientist with your whole being. You have to divide yourself into man and scientist if you don't want to perish mentally. The thought of being a scientist is no longer a happy one.

Science is gradually sinking to the level of a mere bread-and-butter study. Young people come to the universities, young, fresh, with hearts full of life, full of expectations, full of longings, full of hopes of receiving something great there. In the first semester, you enter the lecture halls with a certain amount of shyness, look up at the professors with awe, marvel at all the new and novel things you encounter. By the second semester you are already jaded, indifferent. You hear the same things everywhere. Nowhere is there a real connection to life. In botany, you start with the study of cells; then you go to zoology, where you start with cells again; then you go to anatomy, where you are presented with the human being as a cellular state. And so it goes on. Everything is frayed, torn apart. You're faced with corpses everywhere. In front of plant corpses, animal corpses, human corpses, nowhere in front of real life. But young people need life, a lively, realistic view of nature. Only a few succeed in fighting their way through dead university knowledge to a living understanding of nature and people.

We younger doctors who were lucky enough to find anthroposophy and who have had the privilege of listening to Dr Steiner's lectures for several years are in a position to see the struggle that today's young people have to fight, and we also know that the young people who come after us will have a much harder time than we did. The gap between the young and the older generation is widening. On the one hand, we must now learn to continue to understand the young people who are growing up.

On the other hand, we want to learn grow old in the right spirit, as described in the words to the older members in Newsletter No. 8.

We therefore gratefully welcome the founding of the Section for the Spiritual Striving of Youth and ask to be allowed to help in its realisation, if the Executive Council of the Anthroposophical Society deems us suitable. Since we are now endeavouring, in the spirit of the course we were allowed to receive at Christmas, to work our way through to a real knowledge of the healthy and sick human being, we would like to set ourselves the goal of training ourselves in such a way that we can, on the one hand, convey to the young anthroposophical non-physician the necessary knowledge about his own human being in such a way that he need not have the feeling of facing a scientist who is "not enough of an anthroposophist". On the other hand, we must learn to bring anthroposophy to non-anthroposophical physicians in a way that also honours science. In grateful respect

Dr Hans Altemüller, Lilly Kolisko, Helene Grunelius, Maria Hachez, Dr Julie Bort.[112]

Lilly Kolisko also attended the Easter course for the "young physicians."[113] In a conversation at the end of this course with Ita Wegman and Rudolf Steiner, the question from the letter quoted above was raised again: "How can the physicians integrate themselves into the Section for the Free Spiritual Striving of Youth at the Goetheanum, and what is their task precisely as physicians within it?" Rudolf Steiner replied: "[...] The physician now has a particular opportunity to intimate the scientific and will therefore be able to give the other youth, who are not physicians by destiny, much that can make them more concrete, more soulful. The physician is the first to enter concrete life, so it is possible that he can have an extraordinarily stimulating effect on the other youth." And again Dr Steiner pointed out how it would be good if medical practitioners could familiarise themselves as intensively as possible with anthroposophical education. In ancient times, healing and education were closely related concepts. [114]

Rudolf Steiner had previously spoken in his lectures about the incarnation of the human being and the efficacy of the planets and their connection with the metals, among other things. Lilly Kolisko took notes and subsequently published three detailed papers on the lectures of the Easter Course in the weekly journal *Nachrichtenblatt*.[115] Her shorthand notes later became a great help in publishing the lectures.

In her presentation of 25 May 1924, Lilly Kolisko wrote:

> The whole cosmic development of man passed before the listeners in a marvellously coherent picture. How the Moon regulates the form of man, Saturn the formless spirit, and the Sun the soul. Only from such insights into the human being will it be possible to achieve a recasting of medicine. We will be able to know how lead, silver and the other metals work in the human

being if we can see how Saturn, the Moon and the other planets are involved in the creation of the human form. The entire solar system is connected with the primary healing powers: Saturn – lead, Jupiter – tin, Mars – iron, Sun – gold, Venus – copper, Mercury – mercury, and Moon – silver [...].[116]

Content like this developed into important new research impulses for Lilly Kolisko. During the years 1924 and 1925 she took part in all the important medical courses to which otherwise only doctors were invited, including the Pastoral Medicine Course. As a "non-medical person", she became a member of the Medical Section at the Goetheanum, which was originally only intended for doctors.[117]

8.

The Agriculture Course: Koberwitz 1924

[...] Since Frau Dr. Kolisko's research work on the activity of "smallest entities" so brilliantly established as fact what until then had been more guess-work in homeopathy, we can, I think, regard it as a scientific fact that it is from the small entities (quantities) that the radiating forces necessary for the organic world are released, when these small entitles are used in the appropriate way. [...].[118]

Lilly Kolisko was invited by Rudolf Steiner to the 'Agricultural Course' in Koberwitz from 7 to 16 June 1924. Again, new tasks awaited her. Later, in England, she wrote about a conversation with Rudolf Steiner there:

"You've got enough to do now, haven't you?" "Oh yes," I replied, "I have enough to do," and enumerated the different investigations I was engaged upon. Dr Steiner continued: "I mean the agricultural problems." "But I'm no gardener! I really don't know what I could just do in this direction! Or is there any special experiment to be done?" Dr Steiner smiled in his friendly way and answered my question with another question: "Any special experiment? But of course, *all* the scientific research work has to be done by you. The agricultural people have to do all the practical work, but you have to carry out the scientific researches. These cannot be done by the gardeners or farmers [...]."[119]

The presentations in the Agriculture Course also revealed something completely new for the farmers. Rudolf Steiner wrote to Ita Wegman from Koberwitz on 10 June 1924:

[...] The agricultural lectures seem to be going well for me so far. I approached them with little hope. But now I am succeeding in extracting points of view from the spiritual world that will perhaps be extremely fruitful for practical agriculture. As far as I can see, the things even make sense to farmers, even though they represent something quite alien to current views [...].[120]

During the conference, Rudolf Steiner again mentioned the work of Lilly Kolisko in a lecture in which he spoke about fertilising and the biodynamic preparations:

> [...] Therefore we need to treat our manure not only as I indicated yesterday; we should also subject it to a further treatment. And the point is not merely to add substances to it, with the idea that it needs such and such substances so as to give them to the plants. No, the point is that we should add living forces to it. The living forces are far more important for the plant than the mere substance-forces or substances. Though we might gradually get our soil ever so rich in this or that substance, it would still be of no use for plant-growth, unless by a proper manuring process we endowed the plant itself which the power to receive into its body the influences which the soil contains. This is the point.
>
> The men of our time are altogether unaware how the minutest quantities will often work with great intensity, precisely where living things are concerned. Now, however, we have the brilliant investigations of Frau Dr. Kolisko on the effects of "smallest entities." What hitherto, in homeopathy, was a blind groping in the dark, has here been placed on a sound scientific footing, and as an outcome of her work I think we may take it as proved that in the minute entities, in the minute quantities, the radiant forces we need in the organic world are really set free—provided only that we use these entities in the proper way. And in manuring it is not at all difficult for us to use the minute quantities in the proper way.[121]

Lilly Kolisko took down the Koberwitz lectures and discussions in shorthand, and it is again thanks to her diligence and attention that the Agriculture Course was able to appear in a more authentic form from the second edition onwards. Guenther Wachsmuth wrote in his editor's note:

> [...] In producing this second edition, I would like to thank Mr E. Pfeiffer in particular for his substantial help in reviewing and correcting it, and Frau Dr L. Kolisko

for providing me with her shorthand notes, which
resulted in important text improvements and additions
to the text of the first edition [...] [122]

As she was also present at the karma lectures and Class lessons in Breslau, she also took notes there,[123] as well as at the curative education course in Dornach in June 1924, which she converted into a typescript together with Albrecht Strohschein and Karl Schubert in 1925, and which formed the basis for the later publication.

During the lectures in Koberwitz, Lilly Kolisko discovered many new research tasks, which she worked on in Stuttgart and later in England. For years she conducted experiments on the various aspects of the agricultural course - from planetary influence to composting, from the production and use of biodynamic preparations to the control of weeds and animal pests, including the use of potentised substances, etc. In 1946 she was able to publish the first comprehensive compilation of her work in this field.[124]

Experiments underground

One of the new areas of research, which is not very well known today, that she began to explore after the Agriculture Course was the study of crystallisation forces over the course of the year. These experiments were inspired by the following course statement by Rudolf Steiner in Koberwitz:

> [...] The opposite is true of the water and of the solid earthy element itself. They become still more dead inside the Earth than they are outside it. They lose something of their external life. Yet in this very process they become open to receive the most distant cosmic forces.
>
> The mineral substances must emancipate themselves from what is working immediately above the surface of the Earth, if they wish to be exposed to the most distant cosmic forces. And in our cosmic age they can most easily do so—they can most easily emancipate themselves from the Earth's immediate neighbourhood and come under the influence of the most distant cosmic forces down inside the Earth—in the time between the 15th of January and the 15th of February; in this winter season. The time will come when such things are recognised as exact indications. This

is the season when the strongest formative-forces of crystallisation, the strongest forces of form, can be developed for the mineral substances within the Earth. It is in the middle of the winter. The interior of the Earth then has the property of being least dependent on itself—on its own mineral masses; it comes under the influence of the crystal-forming forces that are there in the wide spaces of the Cosmos.

This then is the situation. Towards the end of January, the mineral substances of the Earth have the greatest longing to become crystalline, and the deeper we go into the Earth, the more they have this longing to become purely crystalline within the "household of Nature." [...].[125]

Lilly Kolisko began her experiments underground in 1924: With the help of Wilhelm Kaiser, a seven-metre-deep shaft was first sunk on the Uhlandshöhe in Stuttgart, on the site of the Waldorf School, and a few years later it was dug even deeper (16 metres). There, in addition to the crystallisation forces, Lilly Kolisko investigated, among other things, how underground silica amplifies the effect of light in the dark. Under the same conditions, the influence of the moon on plant growth below the earth's surface was observed.

Lilly Kolisko described the shaft as follows:

It was about 1½ metres square and one person could just descend – not very comfortably – on a vertical ladder. The natural underground conditions had to be kept undisturbed as much as possible. That meant that light and air had to be excluded as perfectly as possible. At each second metre, the hole was closed again by a wooden cover, while, the surface of the hole was of course kept permanently closed, being bolted by strong wooden bars.

To make quite sure that no influence of light and air could enter, another smaller channel was dug each metre down, parallel to the surface. To make this quite clear: one deep hole was dug vertically, descending 16 metres. At a distance of every metre, horizontal channels were cut in. The distance between the last two channels was 1.5 metres; that means we had 15 channels distributed over 16 metres. These horizontal channels were again bolted with wood against the vertical channel. On opening the first cover for only a few moments, light and air could enter. Descending, we always immediately shut the first cover, and stood on a wooden board offering just enough space to open the horizontal channel into which the various experiments were placed. Standing on the ladder, the floor of the first two metres could be lifted. We descended, shut the cover, and so on, until we reached the eighth sub-division, 16 metres below the surface of the soil.

Each horizontal channel contained a thermometer to record the temperature, and a hygrometer to show the humidity.

These experiments were very difficult to carry out. The arrangement was rather primitive; there was no security against accidents; no fresh air was available during the whole time we had to work below the surface of the soil. In the beginning it took about 3/4 of an hour to complete the descent and ascent. It was interesting to watch the psychological and physiological effect caused by this experiment. Now it was my task to go down with the experiments. Each week one set of plants went down and another set had to be measured, according to the phases of the moon. For some time experiments dealing with the study of the forces of crystallisation were carried out every day [126]

The hole was dug primarily for crystallisation experiments in connection with the statements made by Rudolf Steiner in the Agriculture Course about the crystallisation forces at work in nature. The opportunity was also to be used to study the influence of the moon on plant growth below the earth's surface. The crystallisation experiments were carried out daily over a long period of time. The plant experiments were placed in the pits built for this purpose every week, according to the different phases of the moon. They remained there for 14 days at a time.

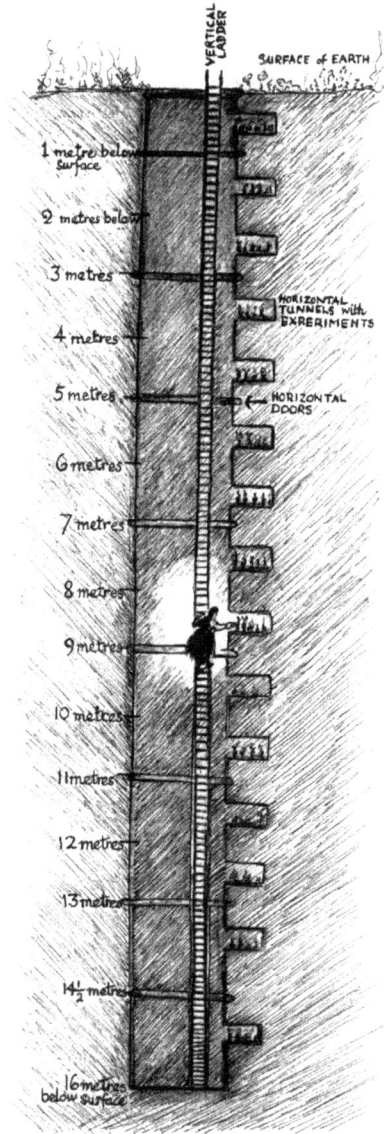

Fig. 20: Drawing by Gladys Knapp.
From the estate of Cecil Reilly England / S. T. Norway

Lilly Kolisko reported what she experienced during the descent into the shaft in her book *The Agriculture of Tomorrow*:

> When I descended, I found the atmosphere damp, and there was a rather mouldy smell; some little creatures, earthworms and centipedes dropped on my head. Suddenly I began to yawn. That was most astonishing. Normally I never yawn, not even after working through several nights! The deeper I went, the more I had to yawn. Then I became aware of my head. I cannot say that my head ached; perhaps I describe my feelings best by saying, I felt that I had a head. I experienced no difficulties with breathing. When I came out again to the surface, I felt a little giddy, there was a strange heaviness in my head and oh! I was tired, tired to death. That is not pleasant, if you have to experience it every day. I remained tired the whole day, and only slowly did my head become clear.
>
> The difference of the seasons is felt very strongly underneath the soil. There are always changes in the atmosphere. For instance, in summertime on descending you feel it is beautifully cool underneath the surface. In Wintertime the opposite happens, it is much warmer than outside. Sometimes I had difficulties with breathing, and one day I simply could not go on.
>
> Something must have been wrong on that special day. I had to find out the reason. The calendar told me it was the 24 December 1931, one day before full moon, nothing else. The next day I went down again to make crystallisations and changed the plants. The same thing happened-only I had made up my mind to go through with it; and, in any case, during the Christmas holidays, I had to work alone. Day after day I had the same experiences - until January the 7th. Quite suddenly the hole was in normal condition again.
>
> I am glad that I went through all these experiences without using artificial breathing apparatus, or any other help, because only thus is it possible to get the complete picture, in uniting the objective and subjective phenomena.[127]

Fig. 21: Lilly Kolisko on her way down the shaft. From Modern Mystic, February 1939

Adalbert Keyserlingk, who was a pupil at the Waldorf School at the time, wrote about these experiments in his memoirs:

> Apart from capillary dynamolysis, Frau Kolisko also worked with artificial crystallization—not the way Ehrenfried Pfeiffer was doing it, but with freely developing, formed-out crystals that could be produced anywhere, even without a laboratory. She had had a 16-metre deep pit dug in which to do crystallizations so that she might show that the etheric presents itself differently below and above ground. One day when she was down in the pit carbon dioxide had accumulated there and she was unable to breathe. She was on the point of fainting when a pupil found her there just in time.[128]

Lilly Kolisko carried out these underground experiments for years. It must have been an enormous test of willpower to prepare the samples with different salts every day for the crystallisations, then take them down to a depth of 16 metres and retrieve them later. She also made comparative samples in the laboratory and in the field. In *The Agriculture of Tomorrow*, she emphasised: "As already mentioned, it is impossible to give a full account here of these numerous experiments, ($72 \times 365 = 26{,}280$ a year)"[129]

Under the earth, she also had to learn to conquer her fear of suffocation. In this way, the deep secrets of the "breathing of the earth" (Steiner) became a living reality for her during the course of the year.

In February 1925, during the last months of Rudolf Steiner's illness, she sent two letters to his studio in the Carpentry workshop in Dornach on the occasion of his 64th birthday, in which she reported on these experiments.

On 16. February 1925, Lilly Kolisko wrote to Ita Wegman:

> Honourable dear Frau Dr Wegman!
>
> Tomorrow is the birthday of our dear Dr Steiner. Here at the Waldorf School, everyone young and old is trying to find something to make Dr Steiner happy. Oh, if only he could see it! How the children sit over their work with burning cheeks and shining eyes because they are allowed to do it [for] Dr Steiner.
>
> I also wanted to get some work done at the research institute so that I could send it to Dr Steiner as a tiny

token. We have been working hard on the crystal experiments, which are to be carried out between 15 January and 15 February. The experiments have turned out beautifully, but it was impossible to finish them by tomorrow. It will still take weeks. Now there is one more piece of work left, which is also taken from the agricultural course and was recently completed. It can be used to show that light in the earth is made more effective by silica. Perhaps Dr Steiner will enjoy receiving this very brief report and looking at the curves. He always enjoyed the graphs when I was able to show them to him in the laboratory.

I am enclosing the report and the accompanying graphs through Miss Rommel and ask you most sincerely, if you think it is good and possible, to give the work to Doctor Steiner. If we remember our leader with special love tomorrow, then we must also remember you, dear Frau Doctor. How can we thank you for preserving Dr Steiner for us and the whole world through your dedicated, tireless and loving care? There are insufficient words with which to thank you. I would like to write to you so often, but somehow I always feel compelled to express the love and deep gratitude with which we all look to you, and when it is written on paper, I realise again that it is nothing, that it cannot be expressed in words. Don't be angry with me for saying it so clumsily! But when people have so often said that they feel towards Dr Steiner like a dear father, to whom they can say anything, in whom they can trust everything - I would like to say: you are like a dear mother to us, who also understands everything that is said to her, and not only that, who also understands what is not said to her.

With gratitude and reverence, your Lilly Kolisko [130]

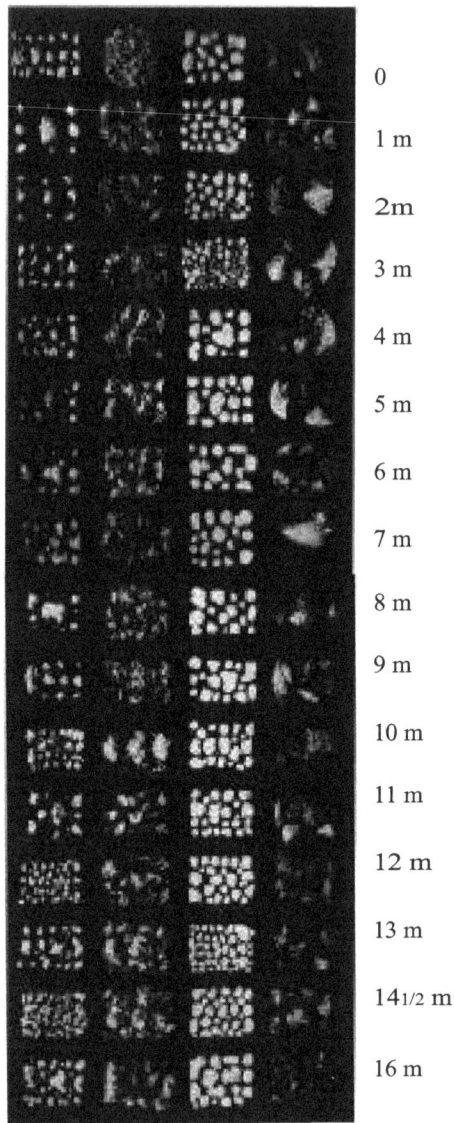

Crystallisation, carried out on 11. January 1931, with lead nitrate, copper sulphate, alum and iron sulphate. 0 = earth surface, followed by the various depths of 1 to 16 metres.

Fig. 22: Crystallisation experiments. © Kolisko Archive, London. From: Lilly Kolisko: "The Agriculture of Tomorrow", page 61

The accompanying letter to Rudolf Steiner read:

> Honourable Herr Dr Steiner!
>
> We've recently tried to complete some projects from the Agriculture Course. For example, from January 15th to February 15th, we conducted crystallization experiments with iron sulphate, copper sulphate, and alum.
>
> If you will allow me, I would like to send you the relevant summary as it stands. I would like to mention that the crystallisation was best at a depth of 1½ to 2½ metres. Deeper into the earth, it decreased again.
>
> A smaller task from the agricultural course: silica brings the light in the earth to greater effectiveness, seems to me to have succeeded well. Allow me to send you a short report with the graphs. I am so pleased to be able to do this work and long with all my heart for the moment when we can show you everything again in the laboratory, dear Herr Doctor. With grateful respect
>
> Your Lilly Kolisko[131]

*

Another question from the agricultural course that inspired her and which she grappled with until the end of her life arose from the following statement by Rudolf Steiner:

> We can see this directly. Look at the green plant-leaves. (Diagram No. 3). The green leaves, in their form and thickness and in their greenness too, carry an earthly element, but they would not be green unless the cosmic force of the Sun were also living in them. And even more so when you come to the coloured flower; therein are living not only the cosmic forces of the Sun, but also the supplementary forces which the Sun-forces receive from the distant planets—Mars, Jupiter and Saturn. In this way we must look at all plant growth. Then, when we contemplate the rose, in its red colour we shall see the forces of Mars. Or when we look at the yellow sunflower—it is not quite rightly so called, it is called so on account of its form; as to its yellowness it should really be named the Jupiter-flower. For the force

of Jupiter, supplementing the cosmic force of the Sun, brings forth the white or yellow colour in the flowers. And when we approach the chicory (*Cichorium Intybus*), we shall divine in the bluish colour the influence of Saturn, supplementing that of the Sun. Thus, we can recognise Mars in the red flower, Jupiter in the yellow or white, Saturn in the blue, while in the green leaf we see essentially the Sun itself. [132]

Adalbert Graf von Keyserlingk was able to witness Lilly Kolisko's sunflower experiments as a pupil at the Stuttgart Forest School because the sunflowers grew on the school grounds:

Frau Kolisko told us that Rudolf Steiner had greatly surprised her by saying that whilst sunflowers always turned towards the sun, they were not dependent on the sun but on Jupiter and should really be called Jupiter flowers. She had tried to influence sunflowers by using several different potencies, and this had proved successful. The next summer we saw an impressive row of sunflowers in front of the institute. The effects of different tin potencies were clearly evident. [133]

Lilly Kolisko extended the connection between planets, metals and plants to the human body organs. And so her research created new starting points for anthroposophical medicine.

*Fig. 23: Sunflowers at the Stuttgart Waldorf School.
From the estate of Cecil Reilly / S. T. Norway*

Fig. 24: Lilly Kolisko in front of the sunflowers. © Kolisko Archive, London

9.

The " last address"

If one has been fortunate enough to be able to come to Dornach again and again and to hear what has been flowing to us from the spiritual world through Dr. Steiner since Christmas, then one feels obligated to bring to the people who cannot be in Dornach at least a tiny fraction of what should now flow as an impulse within everyone. [134]

But if a circle of several or many could be found who would set themselves the first and highest task of allowing those forces to flow to Dr Steiner out of heartfelt love and gratitude, so that he can make use of these forces, perhaps it would be possible him to strengthen his physical powers sooner?[135]

At the last meeting with Rudolf Steiner, his body lay in state in the studio in front of the statue of Christ. Ita Wegman, Count Keyserlingk and Lilly Kolisko observed the wake that night [...].[136]

Thanks to her active participation in the various courses held by Rudolf Steiner, Lilly Kolisko was in Dornach throughout the summer of 1924 until the end of September. She also attended Rudolf Steiner's so-called "last address" on 28 September 1924, shortly before his time of illness.

On 28 September Dr Steiner spoke to the members for the last time. The lecture was shorter than usual, and as you listened you could feel that it was becoming difficult for him to speak. You could also sense more. There was a strange unrest among the people in the Carpentry workshop - where the lecture was held - at least I was clearly aware of it. At a certain point in the lecture, this became even more noticeable. Later I realised what this restlessness was about. There was a murmur among the older members: "What Dr. Steiner said was not true, he was mistaken." I feel it is important to state this here, as it expresses what Dr Steiner had so often pointed out, especially in recent times: the inner opposition, the resistance, which he had to fight against since the founding of the Society in 1912 /13.

I later asked Dr Steiner if I could also bring this lecture to the attention of the members in Stuttgart; he agreed, with one reservation: it must be expressly emphasised that he was unable to complete this lecture. He was too weak. There was still a secret that he had not been able to explain further. I don't want to say any more about it here. Dr Steiner's illness had not come suddenly. It had been clear to those close to him for some time that Dr Steiner's state of health gave cause for concern [...].[137]

The last address touched Lilly Kolisko deeply. During a subsequent trip to Holland at the end of October 1924, she gave an overview of Rudolf Steiner's lectures and the development of the School of Spiritual Science since the Christmas Conference in several cities. She wrote to Ita Wegman about this trip on 30 October 1924:

Honourable Frau Dr Wegman! As you will have seen from my husband's last letter, I had undertaken to accompany Ms v. Mirbach to Holland. I would now like to take the liberty of reporting on my stay in Holland. Immediately after my arrival in The Hague, Dr Zeylmans asked me to give a report to the Hague branch on the recent events in Dornach and also possibly to talk a little about karma issues in general, because the members in The Hague generally know little about them. I wasn't really prepared to give a members' talk in The Hague and I didn't have that much time to ask you again if I could talk about Dr Steiner's last address in The Hague. Now I told myself that the Dutch friends are very far away from Dornach, and if the opportunity is missed now, it might not come again soon.

Dear Dr Wegman! For myself, it is actually the case that since this last admonition, I would like to say, from Doctor Steiner to the members, I can no longer find peace. I feel as if a request has rung through me, just as it once rang in the ears of men: "Go forth and preach ..."

If one has been fortunate enough to be able to come to Dornach again and again and to hear what has been flowing to us from the spiritual world through Dr. Steiner since Christmas, then one feels obligated to bring to the people who cannot be in Dornach at least a tiny fraction of what should now flow as an impulse within everyone. So, I have tried to tell the friends in The Hague what has happened since the moment of the laying of the spiritual Foundation Stone of the new Goetheanum in Dornach. I gave a general overview, as I felt it, and would like to briefly outline what I said.

Firstly, I pointed out the significant fact that Dr Steiner has taken over the leadership of the Anthroposophical Society himself since Christmas, the consequences of this; on the one hand, that anthroposophy now penetrates more into the public sphere, and that those who only become members of the Anthroposophical Society have no obligations other than to be decent human beings. On the other hand, the founding of the School of Spiritual Science. Anyone wishing to become a member of the School of Spiritual Science would now have to take on an extraordinarily important obligation: to become a worthy representative of anthroposophy before the whole world. I spoke about this as my heart dictated. Then I described some parts of Dr Steiner's lectures on karma and gave an outline:

First, about lectures that showed members the way to review their previous earthly lives, and I described in detail the lecture that describes how through the inner activity of the soul, it is possible to form a memory image so strong, so vivid that at night, when the astral body and ego are outside the physical and etheric bodies, it can be taken over by the astral body and elaborated in the world ether. The next day, the astral body imprints it on the etheric body. During the second night, it is further elaborated by the etheric body. The following day, the etheric body imprints it on the physical body. The physical body further elaborates it during the night, and in the morning, the person feels enveloped in a cloud. The mental image descends into the will, the will becomes the soul's eye, one looks back on the causative event in a previous earthly life.

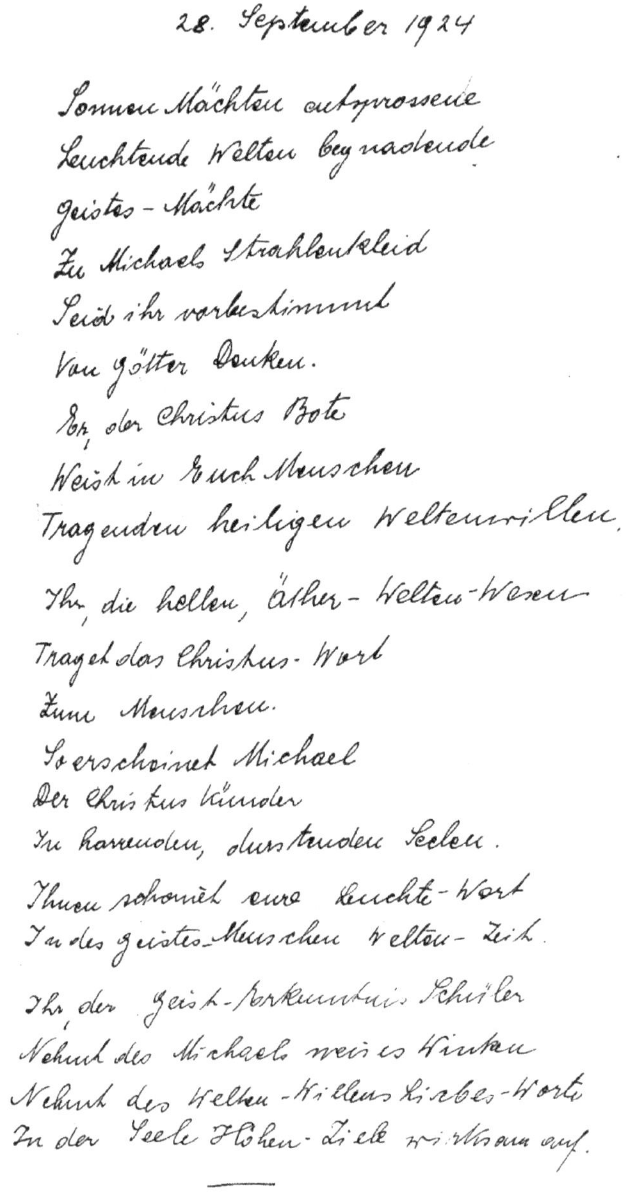

28. September 1924

Sonnen-Mächten entsprossene
Leuchtende Welten bey-gnadende
Geistes-Mächte
Zu Michaels Strahlenkleid
Seid ihr vorbestimmt
Von Götter-Denken.

Er, der Christus-Bote
Weist in Euch Menschen
Tragenden heiligen Weltenwillen.

Ihr, die hellen, Äther-Welten-Wesen
Traget das Christus-Wort
Zum Menschen.
So erscheinet Michael
Der Christus-Künder
In harrenden, durstenden Seelen.
Ihnen schimmert eure Leuchte-Wort
In des geistes-Menschen Welten-Zeit.

Ihr, der Geist-Erkenntnis Schüler
Nehmt des Michaels weises Winken
Nehmt des Welten-Willens Liebes-Worte
In der Seele Hohen-Ziele wirksam auf.

Fig. 25: Lilly Kolisko's meditation book: Meditation spoken by Rudolf Steiner on 28 September 1924, at the end of the last address. © Rudolf Steiner Archive

As a second step, I described the relationship between human beings and their fellow human beings. How pointless it is to observe people as modern anatomy and physiology does, purely externally. A spiritually oriented observation of humanity can only see in a person, right down to their physical corporeality, the result of their moral, soul-spiritual behaviour in previous earthly lives. I elaborated in detail everything that Dr. Steiner said about human physiognomy and gestures. I concluded with a point that one should not only look backward in karmic considerations, but must also direct one's gaze into the future. How it is now, to a certain extent, in the hands of humankind to prepare for their next earthly body, how we will only be able to fight the great battle that awaits us at the end of the century if we possess bodies that enable the higher members of our being to work through our corporeality. I also spoke about the responsibilities we have in raising children, and how immensely grateful we should be to Dr Steiner, who gives us all this and who also gave us Waldorf education.

Thirdly, I moved on to the inner configuration of karma. When a human being passes through the Gate of Death and enters the starry worlds, he enters the lunar sphere and works out his karma there with the great primal teachers of humanity. How a human being must leave behind in the moon everything that is evil in him. He cannot carry evil over into the sun. The sun shines equally on good and evil, but when a human being passes through the Gate of Death, he can only pass over into the solar being with what is good in his being. He passes past the beings of the Mars-Jupiter-Saturn sphere, who examine him to determine what is good in him, and carry what they have perceived down to the lunar beings. On the return journey, a human being must reintegrate into the lunar sphere what he left behind there. We have, therefore, in every human being, a part that has passed through the sun, and a part that has not.

The effect of this fact on the human being's predisposition to health and illness.

Fourth, I referred to the concrete incarnations of individual personalities in history. Here, however, I did not say anything specific.

As a fifth stage, I referred to the Arnhem lectures that the Dutch friends themselves were able to hear from Dr Steiner and which spoke about the karma of the Anthroposophical Society itself.

Having given this foundation, I thought I might venture to give the talk that Dr Steiner gave before St Michael's Day and which speaks so powerfully to the hearts of the members. The lecture lasted about two hours and, I believe, it fell on fertile ground and found receptive hearts.

At the end of my presentation, Mr Timstra asked me to give the same lecture to the old branch of Hague members the next day, as they could no longer be notified for Sunday. I could not resist this request and repeated the lecture the next day.

Now I was asked by the Utrecht friends to come to Utrecht too. I did so on Thursday and found a small circle there too, but a warm atmosphere. The Utrecht friends now asked if I could give a lecture if I was still in The Hague on Sunday, in which case they would come to The Hague and bring along the Amsterdam and Rotterdam members if they could be reached. Now I didn't want to miss any opportunity to unite people in a common endeavour, and so I agreed. On Sunday, I spoke about the lecture Dr. Steiner gave in Dornach immediately before his trip to London, about the task of the Anthroposophical Society to restore the truth of karma, which had fallen into disorder due to the defection of a portion of the Angels from the hosts of Michael, that since a certain point in evolution, there have been celestial and terrestrial angels, the latter of whom now wish to intervene in human development from Earth. Since the Angels are the beings who guide people from one earthly life to another, this fact could not have been without an effect on human destiny. Disorder entered into karma, and cosmic law was disrupted. What unites the members of the Anthroposophical Society? That they can put their karma back in order. Second, I spoke about the last lecture in the karmic series, Plato – Nun Roswitha – Karl Julius Schröer.

Monday morning, I received a telegram from Cologne from Müller-Fürer with a request for a branch lecture for the Cologne friends on my return journey. I then gave roughly the same lecture in Cologne on Tuesday, which was intended to summarise what had happened in the Anthroposophical Society between Christmas and the end of September.

I felt obliged to give you, honourable Frau Doctor, a brief report on all this and hope that you will not be dissatisfied with me for giving these lectures to members without asking in advance. If I had been able to stay in Holland for longer, I would have asked your or Dr Steiner's permission beforehand, but this was not possible. I hope that you can approve of my behaviour [...].[138]

The esoteric youth circle

Lilly Kolisko was very concerned about Rudolf Steiner's state of health and was inwardly preoccupied with the question of how she could help him and how healing powers could flow to him through a circle of people. During her visit to Holland, she discussed this with Willem Zeylmans van Emmichoven and Pieter de Haan - and wrote a letter to Ita Wegman:

> [...] Now it remains for me to say something about more intimate meetings with the Dutch friends. Dr Zeylmans works with great dedication and, I believe, with great success, especially in medical circles. He gives lectures in Utrecht, Amsterdam and Leiden - more and more people are coming. I spoke to Dr Zeylmans late into the night every evening about how we should continue to work in the future. We spent a lot of time discussing Dr Steiner's current state of health and whether there was a way to give him the support he needed. It is clear that the individual person can do nothing in this respect.
>
> But if a circle of several or many could be found who would set themselves the first and highest task of

allowing those forces to flow to Dr Steiner out of heartfelt love and gratitude, so that he can make use of these forces, perhaps it would be possible him to strengthen his physical powers sooner? Oh, dear Dr Wegman, it is terribly difficult to express this. I hope you understand what is actually meant by that. We have all been so worried since we read the latest news in the newsletter, where Herr Dr himself wrote that he could not do anything physically at the moment, that we are looking for a way to let these physical powers flow to him. We thought of the Rosicrucian lectures, of the sacrifice of one part of the circle, so that the other part could be able to do certain things. We would like to make every sacrifice.

Mr de Haan later joined these meetings. - Dr Zeylmans wanted to find people for Holland who would be prepared to work actively and take on the responsibility. They once said that if there were only seven people in each country, it would work. Dr Zeylmans only found Mr v. Bemmelen and his wife for the moment. We asked them to come to us and wanted to talk to them about all the worries and tasks. It took a long time before we understood each other to some extent. These conversations led to Dr Zeylmans and Mr de Haan asking to join the circle to which Mr and Frau v. Bemmelen belonged. Mr v. Bemmelen will report on this himself and forward the request to Dr Steiner through Dr Röschl. However, this conversation did not lead us any further towards the desired goal of Dr Steiner's personality.

We would now like to discuss everything that concerns us with the members of the circle who are in Dornach on 6 November, during which time a meeting of the Medical Section will be held. I hope I can also attend this meeting. Perhaps you will then be able, honourable Frau Doctor, to offer us your advice and support so that we can find the right solution. [139]

On the same day, Dan van Bemmelen also wrote a letter to Maria Röschl about the inclusion of Willem Zeylmans van Emmichoven and Pieter de Haan in the "esoteric youth circle" - van Bemmelen and Röschl were leading figures in it.

Dear Miss Röschl, by now you will probably have received the news that two new members want to join our youth circle at once. It is a great happiness and a great joy for us. We have been waiting for them for a long time [...]. Frau Kolisko has been here for a week and we met up last Sunday. Pieter de Haan was also there. Dr Zeylmans and he talked about Dr Steiner's illness, about the need for a spiritual community around him that will accept Dr Steiner's will, but also has the will itself to realise it through the power of the community. As individuals we are weak, but as a community we can form something like a body of strength. De Haan also said that we have to replace what Dr Steiner is now lacking. A way must be found in such a community to care for him, to replace his physical strength.

I gradually had to tell them that such a community existed. I didn't yet know that Frau Kolisko had been accepted into the circle. This led to misunderstandings at first, and Dr Zeylmans also had difficulties because he belonged to the medical circle. Frau Kolisko then took that out of his way. Their decision to join the circle was thus finalised.

But they still wanted the circle to ask the question: How can we replace in Dr Steiner what he currently lacks? We would like to use this as a stimulus for a discussion in the circle. With best wishes from both of us, also to all friends, Dan van Bemmelen.[140]

Lilly Kolisko's endeavours for Rudolf Steiner's recovery possibly led her to the decision to become a member of the "Youth Circle" founded in 1922.[141] In her meditation book it says before the Youth Circle meditations: "Mantram received I. XI. 1924". She wanted to do something active, to sacrifice herself, to serve Rudolf Steiner and anthroposophy in deep earnest.

On 16 October 1922, Rudolf Steiner spoke at the founding of the youth circle about how the power of the community can help the individual:

With a spiritually appropriate attitude, a peculiar relationship will arise in relation to the spiritual substance formed by the meditations, a relationship of each individual to the whole. This relationship will be able to develop in this way: At certain times and for certain tasks, everything that is developed by the community will be focussed on one individual. He will then be gifted with the entire spiritual substance of the community for his tasks. If the others who belong to the community now correctly understand what is happening, they will look on without envy, indeed with a justified sense of joy, at how everything has been given to one person at this moment. Conversely, this person will not only be able to attribute it to his own virtues or talents if he now succeeds in many things. He will have the awareness that he is essentially working with and from what others have given him. And this will call him to humility and gratitude.[142]

In a letter to Ita Wegman dated 21 November 1924, Willem Zeylmans van Emmichoven also dealt with the idea of sacrifice: [...]

Finally, I would like to come back to our conversation about how Dr Steiner might be helped. What you told us at the time went straight to our hearts and was also discussed in the groups. However, as you probably remember, Frau Kolisko and I also asked whether it would not be possible to help by sacrificing what we ourselves hold most dear and which is probably the only thing of any value to us. You told me at the time that there was no need for this, but that the possibility of helping in this way was perhaps there, even though it was not "appropriate" now. At home, I spoke to de Haan, with whom Frau Kolisko and I had already spoken about this earlier. And from this conversation with de Haan came the following thought: even if it is not necessary now, perhaps wrong, if we were to do something like this, it is possible that after some time, perhaps only after years, such a sacrifice might be necessary. This could best be done by people who have been preparing for a long time to make such a sacrifice, so that they can then do it fully consciously and after careful preparation. There will come a time when Dr Steiner's physical powers will diminish, although of course we hope it will take many, many years. But when that time comes, wouldn't it be possible for some people to sacrifice the best and strongest they have in themselves in order to renew his physical powers. I do not know whether such a thought is permissible. Perhaps there is a lot of hubris to think that Dr Steiner would be so helped by weak mediocre people. But I have always thought that weak and mediocre people can become a force through sacrifice. If this thought is not wrong, then both Frau Kolisko and de Haan and I are ready and so we want to prepare for it in close consultation with you and try to find the people who can help us. (So far, conversations have shown that there was no "understanding" for this among those we spoke to about it. Even to our astonishment, not among many in Dornach). But it is also possible that you say: Do your duty where it falls on you and continue to practice humility. We will do that. But you should know what lives in the three of us, and that you know that we will do our duty, if necessary, with all the strength that is within us, and also with all the strength that we can increasingly rise to. [With] our respectful greetings to Dr. Steiner and to you! Yours, F. W. Zeylmans v. Emmichoven[143]

Rudolf Steiner's death

Rudolf Steiner's health deteriorated during the winter of 1924/25.[144] Marie Steiner was often at the Stuttgart Waldorf School for artistic work during this time, including on 29 March, the day before Dr Steiner's death in the studio. Lilly Kolisko later wrote about it:

> On Sunday evening, the so-called "Piper Evening" took place, followed by a second meeting with Frau Kolisko. This was 29 March. After the evening event, Frau Dr Steiner received a call from Dr Schickler on behalf of Dr Noll, who in turn had been commissioned by Dr Wegman: Dr Steiner was feeling worse again. When Frau Dr Steiner asked whether she should come immediately, Dr Schickler, who had also already informed himself about the eventuality, said: It was just like on 1 January. There was no reason to fear anything serious. Frau Dr. Steiner stayed. Between 4 and 5 a.m., another call was made asking her to come. She called

Herr Leinhas for a "Kommande Tag" car, but couldn't leave until 8 a.m. and came too late.[145]

Rudolf Steiner died on the morning of 30 March 1925 at around 10 o'clock. Ita Wegman, who was present, recorded:

> [...] His departure was like a miracle. As if something was understood, he moved on. It was as if the dice of decision had been cast at the last moment. And when they had been cast, there was no longer any struggle, no attempt to remain on Earth. He gazed calmly ahead for a while, said a few more loving words to me, and consciously closed his eyes and folded his hands.[146]

Lilly Kolisko travelled to Dornach; Gisbert Husemann later reported on the events:

> At the last meeting with Rudolf Steiner, his body lay in state in the studio in front of the statue of Christ. Ita Wegman, Count Keyserlingk and Lilly Kolisko observed the wake that night. She spoke of a "vividly majestic image" of the face, "as alive as if he were beginning to breathe again". "Now all the work had to be done alone, without him, and in full responsibility." Many difficult hours followed.
>
> "If anything becomes too difficult, why don't you turn to the spirit of the Institute, it doesn't have its name for nothing." That was a word of comfort she had received from Rudolf Steiner.[147]

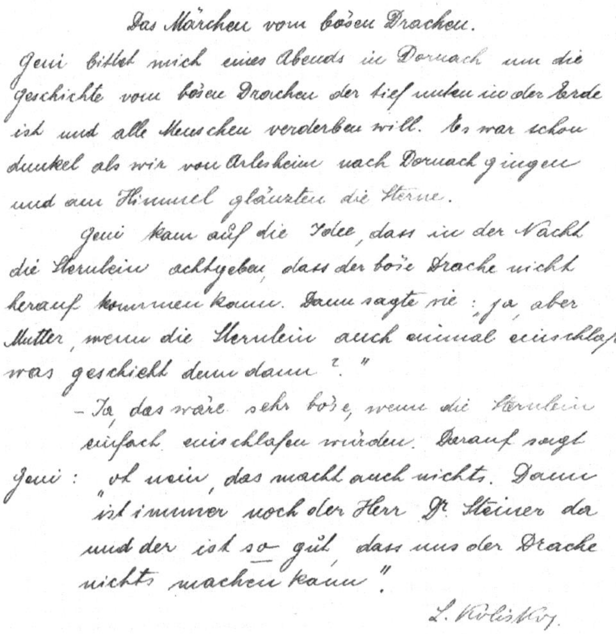

Das Märchen vom bösen Drachen.

Jeni bittet mich eines Abends in Dornach um die Geschichte vom bösen Drachen der tief unten in der Erde ist und alle Menschen verderben will. Es war schon dunkel als wir von Arlesheim nach Dornach gingen und am Himmel glänzten die Sterne.

Jeni kam auf die Idee, dass in der Nacht die Sternlein achtgeben, dass der böse Drache nicht herauf kommen kann. Dann sagte sie: ja, aber Mutter, wenn die Sternlein auch einmal einschlafen, was geschieht denn dann?."

– Ja, das wäre sehr böse, wenn die Sternlein einfach einschlafen würden. Darauf sagt Jeni: oh nein, das macht auch nichts. Dann ist immer noch der Herr Dr. Steiner da und der ist so gut, dass uns der Drache nichts machen kann".

L. Kolisko.

Fig. 26: The fairy tale of the evil dragon. The text is reproduced in the footnote 148 © Rudolf Steiner Archive

115

10.

Marie Steiner

[...] So your wife could soon have her work here, and she is the only one who has testified that she can lead the Esoteric Classes, so she could be considered for this above all. [149]

Thanks to Marie Steiner's regular visits to Stuttgart before Rudolf Steiner's death, Lilly Kolisko's connection with her at this time was very close. Lilly Kolisko therefore asked Marie Steiner in a conversation whether she would like to take part in one of the Stuttgart Class lessons. Frau Dr Steiner agreed, and later Lilly Kolisko wrote the following about it:

[....] This conversation had gone so well that I had gained a great impression of Marie Steiner's personality. Before I said goodbye, she spoke particularly warm words to me: she was completely convinced that I still had a great task to fulfil and that I was destined for something great. When the task was given to me, she was sure that I would also be up to the task... She then attended the Class Lesson I hold on Thursday and assured me afterwards that she had found it very beautiful. It could not have been given better or more beautifully. [150]

Marie Steiner was also present at another Class Lesson: "A second Class Lesson took place in Stuttgart in the presence of Frau Steiner. She first told me that she wanted to attend 'incognito' but then did so publicly in front of all the members." [151]

On 14 March 1925, Lilly Kolisko wrote to Marie Steiner:

Esteemed, dear Frau Doctor! Tomorrow is your birthday. Kindly accept my spring greetings. My heart is full and all I want to do is thank you again and again. I still can't believe that so much love has been given to me. I have received so little love in my life, and basically I walked amongst everyone as a stranger. I wanted to give people everything, but I kept them away from my inner self, from my soul. All the suffering had also filled me with unspeakable bitterness. Now it has all been swept away. I feel blessed by you, dear, good Frau Doctor. Since the first time I was fortunate to see you, my heart urged me to speak to you just once. Your greatness, your majesty closed my mouth. I don't know what gave me the courage this time - but it was an hour of destiny for me. I do not yet deserve your kindness, but I will try to become worthy of it. I feel as if I have finally found my home after a long journey. You spoke a word full of love today - child - oh, if only I could say mother for once! With infinite gratitude Lilly Kolisko.[152]

An indirect response from Marie Steiner can be found in her letter of March 23, 1925, to Rudolf Steiner, seven days before his death:

"[…] Frau Kolisko has become so close to me. I didn't even know that she had longed for this. Now she wants me as her mother, and I must give such a prominent daughter the time she desires."[153]

Lilly Kolisko also took part in Rudolf Steiner's cremation in Basel. She experienced the time that followed as very harrowing.

The funeral took place in Basel on 3 April 1925. A disagreement had already arisen, as Frau Dr Steiner wanted to use the studio to sort the estate, but Frau Dr Wegman wanted to keep it untouched for the members. On the way home from the cremation, there was an open dispute over the urn containing Dr Steiner's ashes in front of the staff at Villa Hansi. Herr Steffen had a cardiac spasm. Frau Dr. Steiner wanted to take the urn directly to Haus Hansi, while the other board members thought it would be taken to the studio. [154]

Stuttgart, 14. März 1925.

Hochverehrte, liebe Frau Doktor!

Morgen ist Ihr Geburtsfest. Nehmen Sie gütigst den Frühlingsgruss entgegen. Mein Herz ist voll und ich möchte Ihnen immer aufs Neue nur danken.

Noch kann ich es nicht fassen, dass mir so viel Liebe gegeben wird. Es ist mir so wenig Liebe im Leben widerfahren und im Grunde genommen schritt ich fremd durch alle Menschen hindurch. Ich wollte den Menschen alles geben, hielt sie aber von meinem Inneren, von meiner Seele fern. Das viele Leid hatte mich auch mit unsäglicher Bitternis erfüllt. Nun ist alles wie weggefegt. Ich fühle mich gesegnet durch Sie, liebe, gute Frau Doktor. Seit dem ersten Mal wo ich das Glück hatte Sie zu sehen, drängte mich mein Herz dazu, nur ein einziges Mal Sie sprechen zu dürfen. Ihre Grösse, Ihre Hoheit, verschlossen mir den Mund. Was mir dies mal den Mut gab, weiss ich nicht – es war aber eine Schicksalsstunde für mich. Noch verdiene ich Ihre Güte nicht, aber ich will versuchen ihrer wert zu werden.

Es ist mir, als hätte ich nun endlich meine Heimat gefunden nach langen Irrfahrten. Sie sprachen ein Wort heute aus voll Liebe – Kind – ach dürfte ich doch einmal Mutter sagen!

In unendlicher Dankbarkeit
Lilly Kolisko

Fig. 27: Letter from Lilly Kolisko to Marie Steiner, March 14, 1925.
© Rudolf Steiner Archive

At the same time, a long-prepared pedagogical conference took place in Stuttgart, at which Miss Mitscher handed over a letter from Frau Dr Steiner to Maria Röschl and one to Eugen Kolisko on the morning of 5 April.
Marie Steiner to Eugen Kolisko:

> Esteemed Dr Kolisko,
>
> I am writing this to you out of great concern for Rudolf Steiner's work. I have clearly recognised that our Executive Council is nothing as it is now orphaned in its infancy. Above all, I am worried about Albert Steffen. It would be our salvation if he were supported by a man who would take on the burdens of the first chairman, who would be a speaker, worldly, energetic, experienced, who had undergone Dr. Steiner's training, and who possessed correctness and tact. That man is you alone. There are never two men in a critical moment for the most responsible position. And for the finest spiritual flow there is now only Steffen. He must be protected from the brutal outside world. I have heard that your presence here has already been discussed and that the Research Institute was to be affiliated to Dornach.[155] So your wife could soon have her work here, and she is the only one who has testified that she can lead the Esoteric Classes, so she would be the one who could be considered for this above all.
>
> I am writing to you privately first. I would be very grateful if you would come here after the conference to discuss this with Steffen. After a night of very careful thought, I realised that this is the only way out. I was unable to get up this morning and therefore couldn't go to the Executive Council meeting. However, I wrote to Mr Steffen with the above in mind and asked him to communicate my conviction with the others.
>
> In my place, I suggested electing Frl Dr Röschl: She will certainly be happy to accept, and it is good that she then works in both societies. There is no possibility for me to remain on the Executive Council; it would not be good. I can serve best by stepping aside now. And I have the section for the performing arts. You might be surprised. Well, Stein has had a vivid dream. It's probably a matter of fate. And if you and Röschl agree, perhaps the work will

be saved, and Steffen too. This has all been considered with full responsibility. Your devoted M.[arie] Steiner.[156]

Eugen Kolisko travelled immediately - without showing the letter to anyone, not even his wife - to Dornach where he was due to attend a medical meeting. But first he wanted to talk to Marie Steiner and the other members of the Executive Council. In several conversations, he firmly rejected his appointment.[157] He wanted the rules of the Christmas Conference to be upheld and the Executive Council appointed by Rudolf Steiner to be confirmed in its task.[158]

Lilly Kolisko wrote about herself and the events over three decades later:

> [...] On 4 April [1925], Frau Kolisko received a visit from an older member from outside - Frau Geelmuyden - who asked if she could read Frau Kolisko's Class lessons, which request was refused. She asked who would continue the esoteric work. Dr Steiner had not appointed a successor. Frau Dr Steiner already had very specific plans, which she would discuss with Frau Kolisko. Frau Geelmuyden could not say anything more specific now. "Well, you'll hear for yourself ..." Now I also travelled to Dornach and was immediately received by Frau Dr Steiner at Villa Hansi. The first thing she said to me was: "I suppose you've heard that you've been asked to hold the Class lessons here in Dornach." I was very upset about this and said I couldn't do that. Surely it should be done by a Executive Council member in Dornach. Wouldn't she like to do it herself.
>
> Frau Dr Steiner refused. Executive Council members would find the words stuck in their throats. They had heard Dr Steiner say these words so often here. Well, I replied, the same applies to me. Frau Dr Steiner replied that she had attended two of my Class lessons herself and that they had been good. Only I was eligible, and I would hear more about them.
>
> A second conversation with Frau Dr. Steiner also took place at Villa Hansi. Frau Dr. Steiner told me about her conversation with Frau Dr Wegman and asked how it was possible for me to have the Class lessons written down. She herself had asked Dr. Steiner if she could have them written down. He then became as angry as she had ever seen him and said: "These Lessons don't exist! They're not there at all!"

But she still had them written down behind the curtain... She did not agree that Frau Dr. Wegman should read the Lessons [...].

She thinks it is wrong that *all* Executive Council members should have the Class lessons [...]. Again, I objected and pointed to the esoteric Council. Frau Dr Steiner: "The Executive Council is *not* esoteric. It was only justified as long as Dr Steiner was alive.

"With his death, it is *nothing...*" She was thinking about expanding the Executive Council. Dr Kolisko for the scientific role, Frau Kolisko for the esoteric side. No one could replace Dr Steiner.

This second conversation had shaken me to the core. I resolved not to speak about it to anyone, not even to Dr Kolisko ... When I returned home in a rather depressed mood, Dr Kolisko asked me whether Frau Dr Steiner had perhaps spoken to me about him. After I briefly answered in the affirmative, he showed me the letter he had received. Only then did I tell him part of the conversation with Frau Dr Steiner. He guided me to go to Dr Wegman and talk to her.

I was reluctant. In the end I agreed to do it. Frau Dr Wegman immediately asked me why I hadn't come to see her. She could see that I was very sad. Not just about Dr Steiner's death, but about something else. Did I not want to tell her? She didn't know whether I still had confidence in her. I only told her part of my conversation with Frau Dr Steiner but left out all the details that I assumed would offend Dr Wegman.

Despite my gentle behaviour, she burst into tears and assured me that she wanted to remain loyal to Dr Steiner. She demanded that I repeat what I had told her to Mr Steffen, Dr Vreede and Dr Wachsmuth. We had to try to confine the matter to this small circle.

The conversation with Herr Steffen, Dr Vreede, Dr Wachsmuth and Dr Wegman took place. I repeated everything. Herr Steffen: "Yes, well, what's next? Now one more person knows. But maybe it really is the case that the Executive Council is no good, so let's resign! Then we should elect a new Executive Council!"

Now they wanted to try to convene an Executive Council meeting at which Dr Kolisko and I would also be present. Frau Dr Steiner was to be asked not to let these things leak beyond the Executive Council. We were well aware that many members had already heard such statements from Frau Dr Steiner - Herr Steffen didn't feel well and said he didn't know whether he could stand the conversation, whether he wouldn't have another cardiac spasm.

The Executive Council meeting took place in the studio. We waited in front of the Carpentry workshop, sitting on the bench, until we were called. After about half an hour, Herr Kellermüller arrived with the message that Dr Wegman had sent word that we could leave. These are the events that took place immediately after Dr Steiner's death, as far as I recall them.

We returned to Stuttgart. It wasn't long before various rumours began to spread among the members [...]. Once again we went to Dornach, this time together with Dr Stein, and asked for a meeting with Frau Dr Steiner so that she could put a stop to these rumours. The other reason for our conversation was that there was a rumour circulating that she would be asked to take over the protectorate of the German National Society ... she said she knew nothing about the rumours that Herr Walter had spread [...]. Before saying goodbye, she took my hand again and said: "And I maintain that you are the only one who can continue to give the Class lessons. You are completely pure, untainted and can make atonement ..."

[...] Looking back today on the further development of the Anthroposophical Society, one must say that the decline began immediately after Dr Steiner's death. Frau Dr. Steiner resented our failure to comply with her wishes: Dr. Kolisko declined to become First Chair, and I myself refused to hold the Class lessons in Dornach. We did everything we could to keep the Executive Council intact, as it had been chosen by Dr. Steiner. Our loyal conduct had sad consequences. Slowly but surely, everything was moving toward a definite end. At that time, we still hoped that things could be put right again. We were silent and breathed a sigh of relief when, on 3 May, in the Newsletter No.18, Volume 1925, the

Executive Council in Dornach issued the following
letter to the members:

"The leadership of the Anthroposophical Society will
continue in the same spirit as Rudolf Steiner indicated
at the Christmas Conference. It was possible to finalise
the reorganisation of the institutions connected with this
course shortly before his death (see Mitteilungsblatt 22
March 1925), but no later information is available that
would give cause for a change in this situation. The
Executive Council he appointed considers it its duty to
remain in its positions and to continue working in the
spirit of Rudolf Steiner, whom it recognizes as its
continued leader.

We therefore request that all questions concerning the
Anthroposophical Society be addressed, as before, to
the Executive Council (Secretariat: Haus Friedwart, 1st
floor); questions concerning matters relating to
individual sections should be addressed to their heads;
applications for admission to the 1st Class of the School
of Spiritual Science should be addressed to the
Secretary, Frau Dr. I. Wegman.

Above all, the Executive Council has set itself the task
of realizing Rudolf Steiner's primary focus, the
construction of the Goetheanum. It counts on the
enthusiastic participation of its members.

Signed: The Executive Council of the General
Anthroposophical Society, Albert Steffen, Marie
Steiner, Dr. I. Wegman, Dr. E. Vreede, Dr. G.
Wachsmuth."[159]

Lilly Kolisko wrote:

As I said, we breathed a sigh of relief that the
Executive Council of the General Anthroposophical
Society had sent this message to the members, that the
Executive Council appointed by "Dr Steiner"
considered it its duty to remain in its functions and to
continue to work in the spirit of Rudolf Steiner, whom
it knew to be a constant leader in its midst. We hoped
that now, after the declaration of the united Executive
Council, matters would have come to a conclusion.[160]

*

Marie Steiner was also frequently in Stuttgart in the following years, and Lilly Kolisko initially endeavoured to stay in contact with her despite the estrangement that had set in. On 25 November 1926 she wrote:

> Esteemed Frau Dr Steiner! I am very sorry that it is not possible for you to see me. Therefore, please allow me to say in writing what I would have preferred to say in person. Every Thursday at 9:30 in the evening I convey the Class lessons given by Dr Steiner to the members of the First Class of the School of Spiritual Science. It is therefore incumbent upon me to do the same today. I would now like to request that the inspection be carried out in the usual way. Mr Reebstein is in the habit of going through the rooms with Mr Geuter at 8:30 a.m., members are admitted at 9 a.m., the building is closed 9:30 a.m. and the bell is switched off. Should you, esteemed Frau Doctor, be disturbed by this and wish to make a change, please inform me in good time of the changes you wish to make so that I can ensure that everything runs smoothly without disturbing you. Expressing the hope that your health will soon improve again, I sign this letter with the most reverent gratitude, Yours, Lilly Kolisko[161]

11.

The working relationship with Ita Wegman.
Questions of destiny and Class lessons

> [...] It is also my conviction that no-one can or should move away
> from the position where Dr Steiner has placed them. Unless his
> inner self speaks convincingly in favour of him having to leave.
> You can't become a deserter now! I, for my part, will remain in
> my position until the last minute and remain loyal to Dr Steiner.
> [...].[162]

After the Christmas Conference, Rudolf Steiner developed two important and concrete ways of working with karmic connections: on the one hand, through the weekly karma lectures held in Dornach - and later also in other places - with the karma exercises contained therein; on the other hand, through the esoteric lessons of the First Class of the School of Spiritual Science.

During 1924, Lilly Kolisko always attended the karma lectures in Dornach and took short-hand notes. Her detailed presentations in 1924, during a trip to Holland, bear witness to the fact that she knew this content well. She also shared them with a group of young people from the Free Anthroposophical Society in Stuttgart. On 10 May 1925, Ernst Lehrs wrote about this in a letter to Ita Wegman:

> [...] Frau Kolisko holds a lecture from her stenographic notes every evening, and the members of this group try to penetrate the understanding of these lectures through their own work, which they distribute among themselves each time, and to connect in their hearts with the essence that Rudolf Steiner gave them.[163]

On 26 April 1925, Lilly Kolisko wrote a long letter to Ita Wegman in which she reported on her studies of historical personalities in this context:

[...] I have now given the life of Sertorius further study, and I have succeeded in finding the passage in Plutarch's story that absolutely reproduces the image of the dream [...].

The following event in his life seems to indicate that Sertorius had connections to the Ephesian mysteries. One day a man brought him a snow-white "hind" as a gift. At first Sertorius took no particular pleasure in it, but later, when the hind was very trusting towards him, he gradually esteemed it as a being of a higher kind, claiming that it was a gift from Artemis and that many secrets were revealed to him through it. In this way he knew how to restrain the minds of the barbarians, and from then on they were more docile because they believed themselves to be guided not by the mind of a stranger but by the deity himself.

Sertorius also founded a school for boys [...]. In any case, the study of this personality is very interesting [...].

I am still looking for suitable books for the study of the Ephesian Mysteries. The Ephesian Mysteries must also have been Michaelic! They are described in one book as international, equally accessible to barbarians and Greeks, freemen and slaves. I found a lot about Heraclitus in "Christianity as a Mystical Fact," but where can I find Heraclitus's contemporaries? [...].[164]

Another letter (dated 10 October 1925):

Honourable dear Frau Dr Wegman! Enclosed is a copy of the "Cratylus" literature he gave you. This is the last copy. Since time is of the essence, I can no longer transcribe it. On a sheet of paper that you must have, there were still handwritten notes by Dr. Stein, which were not written down. It would be very desirable if you could find the original. As far as I remember, it lay with other letters in a cardboard box, first in the old room at the clinic, thus among various other "mail," then it was allowed to make its way over to the new house. Perhaps Mien Viehof will find it.

With warm regards and grateful respect, Lilly Kolisko.
– The last Michael essay was wonderful. I didn't understand some of it and would like to hear more from you sometime.[165]

At the opening of the Second Goetheanum three years later in Michaelmas 1928, Ita Wegman spoke extensively about Cratylus and the Ephesian Mysteries: "[...] The time when these kinds of Mysteries were at the height of their splendour coincided with that epoch characterized by Rudolf Steiner as a Michael Age. This Michael Age began about the year 601 BC and ended about 247 BC. Into this period falls the life of Heraclitus, the Ephesian philosopher. Also living at that time in Ephesus, as a pupil of Heraclitus, but outliving him, was the philosopher Cratylus.

Rudolf Steiner drew my attention to this person Cratylus. He is chiefly known from a dialogue of Plato bearing his name and from a few sentences by Aristotle in his Metaphysics and Rhetoric [...]." [166]

A letter dated 23 October 1925 to Ita Wegman, reveals that Lilly Kolisko also spoke with Walter Johannes Stein, who was a close friend of Eugen Kolisko, about her reflections and experiences in the context of her historical and karmic studies:

> Honourable dear Frau Dr Wegman! It has been an awfully long time since I have seen or heard from you, so I feel the need to write at least once. I had hoped to get something like an answer to my last representations from my husband, but he was so taciturn that he didn't tell me anything about it. The thing is, I've been dreaming something strange almost every day recently. Now it seems that some of these dreams have at least some reality. I'm sure you'll be interested to hear about one of them, as it is in some way connected with my work in the laboratory. It was last Saturday afternoon when I was talking to Dr Stein about "Wallenstein". Perhaps you remember the dream I had in which Dr Steiner said: If you want to find your way back to your earlier lives on earth, then you must study an individuality, as Wallenstein was, etc. Now Stein said that W's constant companion was the astronomer Seni. It was actually Seni who caused W. to conduct this strange kind of campaign, to which Dr Steiner had also drawn my attention. I should think about Seni for once [...].[167]

However, further exchange in this field was made extremely difficult by the critical voices in the Anthroposophical Society and in Dornach, especially against Ita Wegman. [168]

Nevertheless, the collaboration between Ita Wegman and Lilly Kolisko continued around Class lessons. Ita Wegman had begun to hold the Class lessons in Dornach.[169] On 19 December 1925 she wrote to Lilly Kolisko:

> Dear Frau Kolisko, please be good and send me as soon as possible, with someone you know, the lessons given after 22 April 1924. This is the last lecture that I still have and that I received from Frau Dr Steiner. I cannot get from Frau Dr Steiner the 2 missing lectures which I still have to give before the 11 May - at least so far I have not succeeded. She doesn't seem to be able to find these lectures. I have received one of the three lectures that I was missing; there is no ill will here, just a lack of time to look for them. But in order to avoid a standstill, I would like to ask you to send me these two missing lectures from 25 April and 2 May 1924.
>
> We are all looking forward to seeing you. It's just that the work pressing down on me from all sides and all the old memories make my heart heavy.[170]

Lilly Kolisko continued her work with the Class lessons in Stuttgart. On 7 July 1926, she wrote to Ita Wegman again:

> I am hereby sending you the 1. and 3. Class lessons. I need the fourth this Thursday. You must then be so kind as to tell me when you need the fourth lesson and possibly send me the 5. lesson [...] two members then contacted me with a request to forward the verses. I will write to them immediately, telling them they can contact you or Frau Dr. St[einer] or St[effen]. Things like this happen all the time, and small misunderstandings arise [...].[171]

Lilly Kolisko also regularly received enquiries from people who wanted to become class members and passed them on to Ita Wegman.

Her report on the situation from a letter dated 12 October 1926 bears witness to a confidential connection and collaboration - in the midst of the increasingly escalating disputes within the Society:

> Honourable dear Frau Dr Wegman! Yesterday I received your kind letter in which you wrote that I should also travel to Holland. I am very happy to do so, but I

do not yet know how it might work as I have no official business there. So, it will be seen here as my private pleasure, which means I will have to cancel lessons at school and also Class lessons. Miss Hillmann has to look after three children at Kühn's for 5 hours a day and has been taking painting lessons with Miss Geck since 1 October. She will therefore not be able to look after Geni. In the meantime, it is Geni's heart's desire to visit the seaside in Holland. Mr de Haan invited me to go with Geni in the summer, which I didn't accept at the time. I wrote to him today asking if it would be possible to come with Geni now. He also has a little girl, so both could play well. I hope that de Haan won't turn it down, so that things can be sorted out.

Geni longs for you very much. When Dr Schickler came back, she asked him straight away if you hadn't said when she (Geni) could come back. The letter Geni sent you is quite remarkable. It's touching that, despite all the work, you still take the trouble to reply to Geni. I thank you from the bottom of my heart for your concern. I had a bad cold and didn't know how I would manage to speak during the Class lessons on Thursday evening. Now I wanted to conquer the cold with all my willpower. It happened that Thursday afternoon, while I was preparing, I suddenly felt the cold sliding down my throat. It stabbed intensely in my chest, burned in the right side of my back, then skittered down to my left hip. It moved back and forth between my chest and hip a few times and then settled on the hip area. I made it through the Class lesson, somewhat stiffly but with an almost completely clear voice, but I couldn't get up on Friday. I'd already had a fever around midday for a few days. When I couldn't walk, Geni apparently wrote to you right away.

Now it's already better. The pain has moved back up to my back, which means I can walk again. A rash has appeared on my face. I'll be able to speak again by Thursday. The social issue is very troubling. Dr. Röschl brings you a letter from Frau Dr. Steiner. She's now speaking out. The whole atmosphere is very heavy here. When will it finally become completely clear?

Mr. Strakosch is working feverishly. We're sitting on

a volcano that could erupt at any moment. It's difficult to look at all this and refrain from judgment. Every judgment is also a "judging," and we shouldn't judge. God grant that we find the right thing in each of these difficult moments. The destiny of humanity is being decided. Tomorrow there will be a meeting between Dr Röschl, Lehrs, Rath, Schickler and some Waldorf teachers. It will depend on this whether we can continue to speak in the free Anthroposophical Society or not. If only I knew for sure that this is what you want. Dr Schickler has today brought different news than Dr Röschl brought. Is it a compromise again, or should it be done differently now? We can't do things by halves now. Either completely or not at all, otherwise we will do more harm than good.

Issue No. 3 of Natura is very beautiful. Doldinger wrote about it in the "Christengemeinschaft", have you seen it?

Dr v. Heydebrand has written a review of my book and sent it to Mr Steffen. How are you, dear Frau Doctor? Would that we were more capable people so that we could relieve you of more worries! With grateful respect, Your L. Kolisko [172]

Two months later, on 20 December 1926, Lilly Kolisko wrote in her reply to Ita Wegman's invitation to the Christmas and Christmas Eve gathering in Arlesheim:

Esteemed, dear Frau Dr Wegman,

To my great surprise I received such a lovely letter from you today. Of course, we are all in a very difficult situation at the moment. The situation is changing almost from hour to hour. It is also my conviction that no-one can or should move away from the position where Dr Steiner has placed them. Unless his inner self speaks convincingly in favour of him having to leave. You can't become a deserter now! I, for my part, will remain in my position until the last minute and remain loyal to Dr Steiner.

Mr Steffen's drama has just arrived. I have read it and believe that it will lead to conflicts in Germany. Should we say it is a courageous act or ...?

Geni often thinks about Dornach and writes one letter after another. I struggle to keep more letters from coming. I'm enclosing today's edition.

Probably only my husband will come for Christmas. I want to stay here and work, because lately I've had to leave a lot of things undone that should have been done. And who knows how long I'll even have the opportunity to work at the Biological Institute! Frau. Luik wrote me a lovely letter from Brazil. She wrote to you in detail. I wish you a blessed Christmas from the bottom of my heart. With grateful respect, Yours, L. Kolisko [173]

Fig. 28: Ita Wegman and Lilly Kolisko in Holland, summer 1925.
© Ita Wegman Archive

12.

Ongoing research in Stuttgart
1925 - 1927

[...] We believe that the financing can easily be secured if the members within Germany, to whom we are primarily addressing this appeal, become aware of the important fact that the School of Spiritual Science in Dornach needs a Biological Institute and that the preservation of this Institute means collaborating in the reconstruction of the Goetheanum [...] [174]

[...] There are so many details in the Agriculture Course that one could try out. I don't know why people always want to do such playful experiments on their own initiative before they have worked out what Dr Steiner himself has entrusted them with [...]. [175]

[...] Except that Frau Kolisko does very beautiful work and in the "Türmer" is a very pleasing critique of the entities work by a scientist, are rays of hope [...]. [176]

Economic support of the institute

The economic situation in Central Europe was extremely difficult due to high inflation, and the funds of the Biological Institute at the Goetheanum were therefore very meagre. Together with Rudolf Steiner, Lilly Kolisko had formulated an appeal to the members in Germany in 1924 to obtain further financial support for the work of the "Biological Institute at the Goetheanum" via the fund "for the reconstruction of the Goetheanum", but the "appeal" failed to materialise due to Rudolf Steiner's illness.

On 26 April 1925, Lilly Kolisko wrote in a letter to Ita Wegman:

I enclose a copy of a draft for an appeal to the members, which I handed over to Dr Steiner at the time. It was intended to serve as the basis for an appeal that the Executive Council would endorse. [177]

133

"Call to the members!

Much has changed in the Anthroposophical Society since the Christmas Conference in Dornach. These changes have also had an impact on the various institutions that have emerged from within the Anthroposophical Society over the years. It was clear that these institutions would only be productive if they were allowed to work in some direct connection with the leadership of the Anthroposophical Society in Dornach in the future.

Since Christmas 1923, the Biological Department of the Scientific Research Institute 'Der Kommende Tag' has been affiliated not only ideologically but also physically with the Goetheanum, the School of Spiritual Science, in Dornach, by being financially separated from Der Kommende Tag, the joint-stock company for the promotion of economic and spiritual values. This incorporation of the Biological Institute into the property of the Goetheanum, the School of Spiritual Science, was made possible with the help of the fund for the reconstruction of the Goetheanum, which was collected in Germany. Until now, the Biological Institute owed gratitude to the Kommende Tag for financial support. Now this institute has transferred to the Goetheanum, and new ways of financing must be sought. We believe that the sufficient finance can be raised if the members within Germany, to whom we are primarily addressing this appeal, realise the importance of the fact that the School of Spiritual Science in Dornach needs a Biological Institute, and that the preservation of this Institute means working towards the reconstruction of the Goetheanum. The contribution that Germany could make to the rebuilding of our beloved Goetheanum could now be that it assumes the obligation to continue to maintain the Biological Institute in Stuttgart, which is to be regarded as a part of the Goetheanum, and to give Dr Steiner the opportunity to carry out those experiments there that he deems necessary.

We therefore turn to all members in Germany with the request to contribute further funds to the reconstruction of the Goetheanum, if possible, to commit to certain monthly amounts. A sum of approx. 800 marks per

month would have to be raised. Notifications should be
sent to: ..."

Research questions from the Agricultural Course

Lilly Kolisko continued her research into the suggestions of the
agricultural course. For example, she used capillary dynamolysis to
investigate the fertilising value of various animal excretions, including
the urine of all kinds of farm animals, which the anthroposophically
oriented veterinary surgeon Joseph Werr made available to her as test
material from his practice. However, Lilly Kolisko did not want to limit
herself to domestic animals: "Apart from the research which we carried
out, having in mind a purely agricultural viewpoint, we also studied
the excretions of other species in the animal kingdom: camels, bears,
elephants, lions. For this purpose, we had to make friends with the
personnel of zoological gardens, or sometimes we asked permission to
collect the excretion of wild beasts in circuses."[178] The method of
capillary dynamolysis was also used to diagnose animal diseases.

Lilly Kolisko had been investigating the influence of the moon on the
growth of wheat, oats and barley since 1921. In 1926 she carried out
outdoor experiments with maize at different sowing times and concluded
that the seeds sown two days before the full moon germinated very
quickly, developed into healthy and resistant plants and gave a much
higher yield than those sown two days before the new moon.

There were also large differences in cut flowers, carrots and other
vegetables depending on the sowing time.

Rudolf Steiner had also given advice on the control of animal pests. Lilly
Kolisko wrote about her experiments with mice in this regard in the book
Agriculture of Tomorrow:

> Dr. Steiner gave a strange advice. We have to look for
> a certain constellation of the planet Venus, collect the
> skins of mice during this time, and burn them. If we
> scatter these ashes over the field the mice will
> disappear. It is not necessary to burn the whole mouse;
> we only need burn the skin. How can we verify such
> assertions? They seem so very strange to us. [179]

To this end, she wanted to carry out experiments with mice during a
Venus constellation in 1926. But to her great horror, the female mice
killed all the males during the constellation! She wrote:

135

Rudolf Steiner really knew about all these forces, how they work in the plants, in the animals and in the human organism. All his indications are correct. If we burn the seeds of the plants we interfere with the forces of reproduction in plant life. If we burn the whole insect, we interfere with the reproduction in animal life, but then we must take the sun into consideration. The sun must stand at a certain place in the zodiac. If we burn the skins of higher animals, we again interfere with the force of reproduction – then we must see that the planet Venus stands in a certain constellation. That we are interfering with the forces of reproduction is quite obvious from the fact that the female mouses killed the male ones.[180]

Lilly Kolisko also received many questions concerning Steiner's suggestions for agriculture. In a letter to Hilma Walter, a doctor from Arlesheim, she explained:

Dear Frl Dr. Walter!
Forgive me for not answering your inquiry immediately, but I didn't know what it was about. Now my husband brought me the letter from Mr. v. Brederlow so I could find out more details. Mr. v. Brederlow approached me some time ago during a train ride with the question of what I thought about treating grapes with a potency of gold. I asked him in amazement why he thought of that. I received a somewhat mystical-sounding answer that he just imagined it: grapes need a lot of sun to thrive, and if you also used gold for fertilizing, that should be good. What potency I would recommend? I then told him that I would use approximately a 12th potency and gave him precise information about the preparation of the potencies.
His next question was, how will I then determine the effect of the gold? I told him that he would have to see what the grapes that were treated with gold looked like compared with those that were not. Now he said that he imagined that if he held the grapes up to the light, he would
see that they were brighter, that you could recognise the gold by the colour of the bright yellow, etc. I was a bit embarrassed by the conversation on the moving train,

where people were always going back and forth. I then
went on to say that you might be able to recognise a
subtle change when eating the grapes if you were able
to perceive it.

He then asked what effect this would have on the
person who ate the grapes. My opinion was that grapes
fertilised in this way would gradually become a
medicine.

That would be one thing. Now I believe, dear doctor,
that in this case you really have to be very careful that
the whole praiseworthy activity doesn't degenerate into
playful experimentation. It is certainly true that a crop
that undergoes gold fertilization can be viewed as giving a
person gold in small doses, gold that has already passed
through the plant and has been further potentised by the
plant. Dr Steiner did not provide anything in the
agricultural course regarding the further development of
viticulture. He did respond to a question about the control
of phylloxera. I would think that for a specific medicinal
purpose, one could have a wine produced that contains
potentised gold, but in general, one should not do such a
thing.

Then the question of lead. Here, Mr. von B. also asked
me what I thought about using lead to combat bacteria and
plant pests. I replied that Dr Steiner had indeed provided
corresponding remedies in the agricultural course. Why
then does he not want to use it? You can't just arbitrarily
inject various metals into the soil. But it seems that Mr. v.
B. wants to do it anyway. The lead idea is based on the
following: In Koberwitz, beautiful roses bloom in the
countess's garden, but for some inexplicable reason
they have turned brown. She turned to Dr Steiner for
advice and he told her: "Yes, the soil here is very rich
in iron and that is too much for the roses. I would advise
you to water the roses with a high dilution of lead. So
here is a specific statement by Dr Steiner for a very
specific case. Lead is prescribed to paralyse the
excessive amount of iron. Incidentally, I was very
surprised by this; I had rather thought that copper
should be given in such a case. Now Mr v. Brederlow,
who probably heard this remark from the countess,
wants to use lead to combat some kind of plant disease

in Auggen, where the soil may not contain much iron at all.

There are so many details in the agricultural course that you could try out, I don't know why people always want to do such playful experiments on their own initiative before they have worked out what Dr Steiner himself entrusted to us.

I enclose the letter from Mr v. Brederlow herewith. I hope that the information I have now been able to give you is sufficient for you to reply to the gentleman accordingly. Yours sincerely, L. Kolisko.[181]

The appendix to the lectures of the agricultural course reads:

[...] During his walk through the flower garden in Pentecost 1924, Dr Steiner looked at the flowers and remarked that they were not feeling well here, there was too much iron in the soil. When he came to the roses, which were flowering badly and did not look healthy (mildew), Dr Steiner advised her to add finely dispersed lead to the soil [...].[182]

The "fertilising" with metals suggested in the last letter to Walter was therefore nothing new to Lilly Kolisko.

In April 1921 Rudolf Steiner had also spoken about such a production of remedies in lectures for doctors and medical students in the presence of Lilly Kolisko and Hilma Walter:

[...] So, suppose we try mineral-metallic remedies. We will then be able to easily understand what we have already learned about the effects of plants. But we will still be able to say to ourselves: Something has happened with the mineral, in that the mineral has already continued into the plant process. And what has happened in the mineralization and vegetative process is a transformation of the mineral forces. So, suppose we set up a sanatorium, surround it with land, and fertilize the land with various minerals. We then allow the soil to be effective, the actual contents of which are open to our knowledge. We cultivate various plants there, of which we say to ourselves: We are using root, herb, fruit, and so on. Thus, we have the process in our own hands, which consists in the plant transforming the mineral into a remedy. One can then intensify this by allowing such plants to develop; one can then treat them as such plants in the way we have just discussed. That is what we want to do at our Stuttgart institute, on the one hand; that is how it must be set up. But one can then go even further. One can now use what one has already obtained from the plant itself as a medicinal product, which can then be used as a fertilizer, and then intensify the force even futher[...]. [183]

Lilly Kolisko commented on this with the words: "Dr Steiner has thus inaugurated a potentiation process that takes place in the living plant organism. A completely new process is introduced into the production of remedies. Rudolf Steiner's detailed instructions were carried out at the Weleda research institute mentioned above. A series of new remedies were created, which were then made available to me so that they could in turn be studied further with the help of the potentiation experiments."[184]

Lilly Kolisko's research was also important for the further spread of anthroposophical medicines. In October 1925, Eugen Kolisko and his colleague Eberhard Schickler from Stuttgart endeavoured to obtain a licence for anthroposophical remedies in Austria. Lilly Kolisko's publication on the "Smallest Entities" was attached to the memorandum sent to Vienna for this purpose[185].

In 1926 Lilly Kolisko published a further book on her potentiation work, which included all seven main metals; she had already written the preface in the summer of 1924 and received permission to print from Rudolf Steiner.[186] This publication was delayed by Rudolf Steiner's illness and death. Ita Wegman helped Lilly Kolisko with the practical aspects of publishing the book.[187]

Her publications received quite positive feedback. Eberhard Schickler wrote to Margarete Bockholt in February 1926: "[...]Exept are that Frau Kolisko makes beautiful work and that there is a very pleasing review of the entities work by a scientist in the 'Türmer' are the rays of hope [...]."[188] And in another letter in March 1926: "[...] There was a review of Frau Kolisko's book in the 'Türmer'. I wrote a short essay with excerpts from it and sent it to Steffen so that he could have it printed in the Goetheanum or in the newsletter. Frau Kolisko could not do this herself [...]."[189]

Correspondence with Guenther Wachsmuth
in the Natural Sciences Section

Lilly Kolisko's research work was not only important for anthroposophical-medical work, but also for scientific and especially biodynamic work. During these years, Lilly Kolisko was in close contact and exchanged experiences with Guenther Wachsmuth within the Natural Science Section.

A letter to Guenther Wachsmuth reveals that she was a member of the Experimental Circle and that she had also been asked by Guenther Wachsmuth to write articles for the Yearbook of the Natural Science Section, "Gäa-Sophia":

> Dear Dr Wachsmuth,
> Enclosed I am sending you an article for Gäa-Sophia. I don't know whether it will suit you. You wanted to know something about the way I experiment, and so I have fulfilled your wish. If you don't like the article, feel free to bin it.
> Yesterday I received four so-called 'Declaration of Commitments' from Dornach. I must confess that I've never seen anything like it. Either one is a decent chap, in which case one doesn't need this complicated declaration, or one is a scoundrel, in which case one doesn't bother. Of course, I still intend to keep quiet about all the information that Dr Steiner provided in his Agriculture Course but with the best will in the world I cannot fill in these declarations.
> Furthermore, I have been sent the wrong copies, as I belong to the Agricultural Experimental Circle and not the working group. I therefore take the liberty of returning the declarations enclosed.
> Wouldn't it be possible for me to get the Bee-lectures (for 12 marks of course), but without the help of a lawyer? I don't understand all this. Dr Steiner gave these lectures to the workers, most of whom are not members of the Anthroposophical Society, and now the members are only supposed to receive them in return for a payment. I would be grateful for some clarification.
> Yours sincerely, L. Kolisko[190]

One of her articles in the "Gäa Sophia" yearbook in 1926 dealt with the collaboration with Rudolf Steiner in the fight against foot-and-mouth disease.

Later, she also wrote about how she became a member of the Experimental Circle:

> "It was during the agricultural course in Koberwitz in 1924. At that time Mr Stegemann asked me to become *an honorary member of the Experimental Circle.* I informed Dr Steiner of this and pointed out that he had said that there would be no honorary positions within the Anthroposophical Society in the future. He replied: 'You can accept that, because you will also have to do the work."[191]

On 20 November 1926, Lilly Kolisko sent an interesting letter to Guenther Wachsmuth about her experiences with the capillary dynamolysis pictures in connection with "ether geography":

> Honourable, dear Dr. Wachsmuth!
> Yesterday I received your beautiful reprint of Etheric Formative Forces in Cosmos, Earth, and Man. Thank you for the kind dedication, but could you please tell me what you would like to emphasize? There's an awful lot on pages 130–245! Should I elaborate on anything, or do you just think I should read all of it?
> What did you think of the silver-iron picture in issue 3 of Natura? It seems to have made an impression on many members. You may also be interested in the following. On my last trip to Holland, I took various substances with me to do some experiments. It is clear that the pictures that emerge on the filter paper must be different in Stuttgart than in Dornach and different again in Holland. After all, they are created through the interaction of light, air and water. They are actually much lighter and softer in Holland than those produced at the same time in Stuttgart. If you could travel the whole world, you would find a beautiful ether geography. When will your second volume be published? Thank you again for sending it and best wishes! L. Kolisko[192]

Guenther Wachsmuth replied shortly afterwards, on 24 November 1926.

Dear Frau Kolisko, I have included pages 130 and 245 because the footnotes refer to your writings on these pages. Perhaps you missed them.

With a heavy heart, I am enclosing another letter from Dreidax, who keeps urging me to tell him what he is doing wrong. Perhaps you would be so kind as to write him a few lines and that would settle the matter.

I was extremely interested in your silver/iron illustrations, and I think the idea of carrying out parallel experiments in different countries is excellent. We increasingly have to relate each individual natural process to the entirety of the earth. This means we are entering an area from the outset that is still largely uncharted territory for science. My ideal is also to gather people in the various countries who are well educated in anthroposophical science over time, so that anyone who wants to carry out parallel experiments in any field in other countries can take contact with the section and will be able to do so.

I believe that this system can be expanded in a few years' time. When I climbed into Vesuvius, I made some extremely interesting observations and brought back some lavas that I would like to show you when you are here.[193]

Lilly Kolisko was invited to the farmers' conference in Dornach, where she lectured on her sowing experiments before the full and new moon, and she wrote another article for "Gäa Sophia" in 1927. Her correspondence with Wachsmuth also continued; she wrote to him in September 1927:

Dear Dr Wachsmuth!

Firstly, I would like to thank you very much for sending me the Bee-course. There was no invoice enclosed, so I don't know how much I have to pay for it. Perhaps you would be kind enough to let me know.

Today I received a request to write an article for Gäa-Sophia. When is the deadline for the article? If the yearbook is to be published at the end of this year, time is terribly short and I don't know how I'm going to manage it, as I have to write another piece of work before Christmas.

Then I would have liked to know whether, under this

article, you mean the publication of the experiments that I reported on during the last farmers' conference in Dornach. That would be the sowing experiments two days before the full moon and new moon as well as the yearly curve on the effects of the moon. The plant pictures will probably not cause such great costs; on the other hand, I could only imagine a publication of the annual curves being valuable if you were able to print the curve in a decent size and with the colours (blue-red, brown, violet, green). If the curve is not clear enough, farmers can't do anything with it. It would be impossible to write about the sowing trials alone. One project without the other is too thin. The farmers were very interested in it last year and would have liked to have had the opportunity to study the curves. I would be very grateful for a prompt reply. Yours sincerely, L. Kolisko[194]

Six weeks later, at the end of October 1927, her next letter arrived at the Goetheanum:

Dear Mr Wachsmuth!
Today, to my great dismay, I received your letter in which you demand the article for Gäa Sophia by 5 November. In your letter of 12. October you gave me a deadline of 10 November. Unfortunately, it turns out that I have to give main lessons at the Waldorf School from 10 October to 12 November in addition to my usual subject lessons, which makes 3 lessons a week. That's the same time commitment as a full-time Waldorf teacher with nothing else to do. I have this enormous teaching load on top of my other laboratory work because the school has no alternative. I'm sure you understand that I find it difficult to find time for another assignment that has to be written responsibly. I will do what I can to meet the deadline of 10 November, I certainly can't do it by the 5th.
Yours sincerely, L. Kolisko[195]

In addition to her research work, Lilly Kolisko also taught at the Waldorf School. A notebook[196] contains entries on her lessons at the upper school. She began the lessons with the "history of writing" through various cultural epochs and types of writing from pictographic writing to runes and hieroglyphics. She also taught shorthand and, in later years, natural history. According to reports by Adalbert von Keyserlingk, the students were allowed to follow her work in the laboratory.

Natura

Natura, mentioned in Lilly Kolisko's letters, was a journal published by Ita Wegman for the Medical Section of the School of Spiritual Science at the Goetheanum to further develop and publish Rudolf Steiner's medical impulses.[197] Initially, the editorial staff was led by Ita Wegman, Eugen Kolisko, Willem Zeylmans v. Emmichoven, Hilma Walter, Margarete Bockholt, and Eberhard Schickler, all of whom belonged to the so-called "esoteric core" of the section, an inner working body that had been formed in September 1924.[198]

In the first volume of "Natura" 1926 /27 there were already several articles by Lilly Kolisko - "On the Mystery of Matter" - in which she first developed the basics of potentiation processes:

> "[...] Here we are indeed faced with the mystery of matter if we want to grasp the concept of potency. We call the rhythmic dilution of matter a potency, which is a *force*. We can push the dilution so far that no chemical or physical means can detect the 'substance'. But in plant growth, for example, we can still show the 'effect' of the potency [...]."[199]

She went on to write about substance, matter, rhythm and force. As a practical starting point for the first two articles, she took the investigations carried out during the foot-and-mouth epidemic, in particular on the shaking time of potentised substances.[200]

In the third article she then reported on the potentiation trials carried out in 1923 and 1926 with peony, calendula and mullein.[201] Summarising, she wrote:

> "[...] These four trials were selected from a large number as typical examples. They could be multiplied at will. Surveyed collectively, one consistently finds three main stages of substance throughout: *sensitive chaos*, *harmony* and renewed *chaos* at a different level. In the three experiments (peony, calendula, mullein) that appear on

the narrow strips of filter paper, you can still see the rhythmic repetition of the banding in the second row, which disappears again in the third row [...]."

In another article she reported on an interesting experiment with red and yellow roses in which the original substance was used as the medium for potentiation ("potentiation in itself" instead of in distilled water or alcohol) in order to make the force visible, which remains invisible in the "normal" potentiation because the proportion of coloured substance parts reduces. In conclusion, she wrote: "Thus we withdraw one veil after another from the face of matter. Its mysterious nature becomes ever clearer to us, and perhaps one day it will stand before us unveiled in all its inherent beauty."[202]

An important and beautiful article on iron and silver also appeared at Michaelmas 1926 with a consideration of the Michaelmas period and the impact of iron and silver in the human organism[203]

An issue of "Natura" in 1927 was explicitly dedicated to Rudolf Steiner to mark the second anniversary of his death.[204] Lilly Kolisko wrote in her article "Memories of Rudolf Steiner" about the origins of her work and its promotion:

"[...] It was wonderful to be able to talk to Dr Steiner about the cell, about blood, about fertilisation. He always led from the smallest things out into the cosmos. From cell life to the starry worlds. Each of his words meant another task to be solved. They were paths from the knowledge of nature to the knowledge of spirit [...]."[205]

Calendula (marigold)

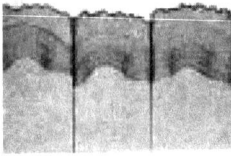

Unshaken   Shaken for a ½ minute   Shaken 1 minute

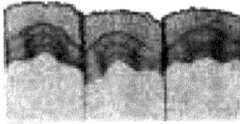

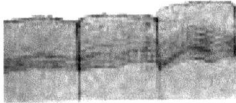

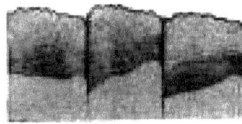

Shaken for 1 ½ minutes   Shaken 2 minutes   Shaken 2 ½ minutes

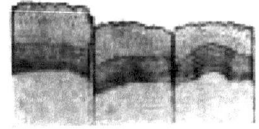

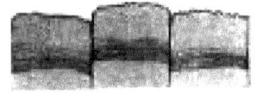

Shaken for 3 minutes   Shaken 3 ½ minutes   Shaken 4 minutes

Fig. 29: Calendula (marigold). © Ita Wegman Archive, from Natura 1926 /27,
Illustrations from pages 180 - 181

13.

The work with Capillary Dynamolysis

[...] Really everything that is done in the [Biological] Institute [at the Goetheanum] is simultaneously a language that says: Man, know yourself! The inwardness, purity and power of the spiritual essences of the metals and elements is also shown unreservedly in the results of the experiments [...] [206]

[...] It is a scientific work that speaks to the artist in man, who can stand in awe before a work of art that shows him the workings of the stars in earthly substances. Rudolf Steiner, the founder of anthroposophy, taught us to look at the world of minerals, plants and animals with different eyes. He showed mankind the way back from matter to spirit. This work has emerged from the inspiration that anthroposophy can provide [...].[207]

[...] Many years of occupation with the many pictures of gold, silver, etc. suggested to me that I should also think of a therapeutic use for these marvellous pictures [...] [208]

From 1926 to 1928, Lilly Kolisko worked intensively with capillary dynamolysis, which she used to analyse metals over the course of the year and during particular constellations of planets.

In this context, Joachim Schultz became acquainted with Lilly's work whilst a student at the Technical University in Stuttgart. Other young people were also very interested, including Adolf Ammerschläger, Franz Lippert, and Ulrich Rudolf. Joachim Schultz described the atmosphere in the "Biological Institute": "Truly, the Goetheanum is at work here! One comes in and is seized by the sustaining spiritual current that creates and works within one. One feels completely addressed and inspired as a moral person. One wants to improve. And one's heart ignites with a luminous glow of sharing in the truest sense."[209]

In his diary entry from Sunday, 18 June 1927, Schultz described further experiences:

[...] Through anthroposophical research work, one becomes much more attentive to these abysses of the human soul. They begin to become more obvious. And it is difficult to bear this. But you must learn to face them in full truth.

Really everything that is done in the [Biological] Institute [at the Goetheanum] becomes at the same time a language that clearly says: Man, know thyself! The inwardness, purity and power that can be brought to bear on the spiritual beings of the metals and elements is also revealed unreservedly in the results of the experiments. The questions that are asked of the cosmos and nature in the experiment are answered according to the degree of maturity of the person who asks and prepares them. Nature does not open up to everyone equally willingly. It cannot be forced.

For example, the same basic phenomenon must always appear in the metal pictures, but within limits. But the heads of elemental beings, the arrowheads, like to appear because they want to reveal something to people. They like to be active, create form after form, can't do enough in harmonious arrangements of contours, in broad, cosy and yet clear and internally radiant arrowheads. You can see this in Frau Kolisko's pictures.

One feels an abundance; a whole stream of beings has been at work therein. For the first time last night I clearly experienced that my pictures are not like those "painted" through Frau Kolisko. Such big differences! Until then, I had almost always found my pictures to be quite acceptable and beautiful. Sometimes Frau Kolisko had also said: "Today you have really nice pictures!" Now I see how I threatened to lull myself into self-satisfied complacency.

When few or only tiny shapes appeared, I naturally attributed it all to the weather, temperature, and other external changes, but not simultaneously to changes in my inner self, to my state of mind and moral disposition [...]. Yesterday, I saw very clearly the opaque beauty of my pictures. They could be 100 times more beautiful. Few forms (relatively), partly encrusted, trapped in dark contours, shaggy - dry, drooping contours. The copper arrows are pointed, cold, and very small. The lead is blurred, without the light-emitting halos like Frau Kolisko's. Frau Kolisko just smiles, looks at both, and says: "Don't you have a relationship with copper?"[...][210]

Lilly Kolisko's capillary dynamolysis work with various metal salts during changing seasons and constellations led to the publication "Sternenwirken in Erdenstoffen"[Workings of the Stars in Earthly Substances] [211] in the spring of 1927. In this study, the connection between lead and Saturn was revealed during a conjunction of the Sun and Saturn on 21 November 1926. The experiments were carried out with silver nitrate, iron sulphate and lead nitrate. Lilly Kolisko wrote in her introduction:

The mystery of matter will be written about here. A new path is being trodden into a vast realm. Wonderful insights and prospects will open up if we look at the signs without prejudice, if we take in the pictures that nature itself has painted. It is scientific work but it speaks to the artist in man, who may step in reverent awe before a work of art that shows him *the workings of the stars in earthly substances.* Rudolf Steiner, the founder of anthroposophy, taught us to look at the world of minerals, plants and animals with different eyes. He showed mankind the way back from matter to spirit. This work is the result of the inspiration that anthroposophy can provide. Every material thing around has weight, suffers gravity, and is subject to the forces of the earth.

Is this weighty body exposed *only* to the forces of the earth, or are extraterrestrial forces also active within it? Can the stars also influence earthly substances? These are questions of immense significance. It is difficult for modern man to believe that a star that is many light years away from the earth can have an effect on the earth. What effect can it have? Light and heat? Both are out of the question at such an enormous distance. This leaves us with invisible forces that cannot be detected by scientific apparatus. And yet the stars influence earthly events! [212]

Lilly Kolisko had very precise ideas about how this publication should be designed within the "Natura series" so that "the outside corresponds to the inside". She wrote to Ita Wegman on 14 June 1927:

> Honourable Frau Dr Wegman!
> I am pleased that you are satisfied with the latest work and thank you for your kind letter. I would like to finish the title drawing, but I find it impossible using "Blue with Gold". The book is about silver, iron and lead. Saturn requires black, the moon white, the iron red. Blue with gold does not correspond to the content, and I tried to harmonise the outside with the inside in the drawing. If there are special reasons for making the title page "blue-gold", then I would suggest printing only the title and omitting any drawing.
> You could also try using a deep dark blue background and printing the drawing in silver-grey if black and white is really too "sensational". But as I said, my suggestion stems from the idea of matching the outside to the inside. Perhaps you will decide in favour of one of these options.
> I saw from Kirchner's letter to Herr Dr v. Grunelius that the format should be smaller. That can't be changed either, because the format depends on the pictures amongst the text, and so I drew the title page very tightly. I showed Herr v. Grunelius exactly how to do this using the pictures [...].[213]

The second publication in the intended series, which had the overall title "Sternenwirken in Erdenstoffen"[The workings of the Stars in Earthly Substances], was published in autumn 1927. It dealt with the solar eclipse of 29 June 1927 and contained pictures with gold chloride, silver nitrate and tin chloride. Both works showed clear changes in the metal salts during the constellations and confirmed the connection between metals and planets.[214]

Lilly Kolisko continued to gain new insights through her work with capillary dynamolysis. In her letter to Ita Wegman in July 1927, she wrote for the first time about the possibilities of using the pictures obtained for therapeutic purposes:

Esteemed Frau Doctor!

Many years of working with the many pictures of gold, silver etc. suggested to me to think about using these marvellous pictures in therapy. I could well imagine that a patient who has a heart condition and is being treated with gold would experience a very beneficial effect from a "gold picture".

Or a sick person who is prescribed lead to draw him into his body would receive strong help through the meditative contemplation of a silver-iron-lead picture.

Various pictures could be reproduced photographically. It is more difficult for gold. Colour would have to be involved here. However, the best colour reproduction will not replace the original especially if it is for a therapeutic purpose. The originals darken in the light. I would now like to try enclosing the picture in glass, but it should only be hung up when you are really looking at it, otherwise it should be covered. I don't know how long the original will last under those conditions. Perhaps 1 to 2 months, maybe even longer, with careful handling? Under no circumstances should it be exposed to direct sunlight. It must hang in the shade. I am sending you the first attempt for assessment through Dr Suchantke. The picture is double-sided. On one side is an original gold picture, on the other a photograph of gold and iron from Easter Sunday 1927. The second gold picture has a photograph of gold-tin on the other side from 14 December 1926.

Perhaps the photographic image belonging to the original could also be attached to the front. But I didn't have any matching pictures to hand at the time. The photograph should always hang on the front and only at times when you look at the original should you turn it round.

The production costs for such a double picture would around 10 marks. Each picture would of course be different.

Mr Suchantke will also bring you photographs of silver and iron with synthetic, alcoholic and digested formic acid. I would like to have these pictures returned to me.

Regarding the gold pictures, I would like to point out that

the dark gold picture would look best in a light yellow-orange-coloured room, while the light picture (which will deteriorate much more quickly) would look best in a violet-pink-coloured room.

With grateful respect, L. Kolisko

P.S. The execution would also be done much more carefully. The pictures were made in great haste so that Mr Suchantke could take them with him. [215]

Whether and how Ita Wegman responded to this suggestion is not known.

Another area of application of capillary dynamolysis was experiments with metal salts in the annual cycle, especially at Christmas and other annual festivals. Lilly Kolisko was invited by Ita Wegman to come to Arlesheim for Christmas 1927 but wanted to stay in Stuttgart for her experiments.

On 19 December 1927 Wegman wrote:

Dear Frau Kolisko,

When your husband was in Brussels, I asked him to ask you whether you would also like to tell the doctors something about your metal experiments. I would now like to ask you personally to help bring some scientific input on your part. This doctors' meeting is intended more as a kind of discussion where the doctors should be given the opportunity to talk about their work, so that everyone will have the opportunity to speak if they wish to do so.

At Christmas, of course, and in general, you are always very welcome here with your husband and Geni, and we will all be delighted to see you here with us again. How are you otherwise? [216]

Eugen Kolisko was also invited to Arlesheim for a medical lecture. His reply stated:

"[...] My wife absolutely wants to carry out the experiments during the thirteen nights here [in Stuttgart] this year and will therefore not be able to leave. She will write to you about this, also because of your suggestion that she should speak to the doctors and physicians [...] [217]

In her experiments, Lilly Kolisko discovered that the pictures change during the major annual festivals and that during the Christmas season, from Christmas Eve to Epiphany (6 January), they depict a special process that reflects the entire cycle of the year.

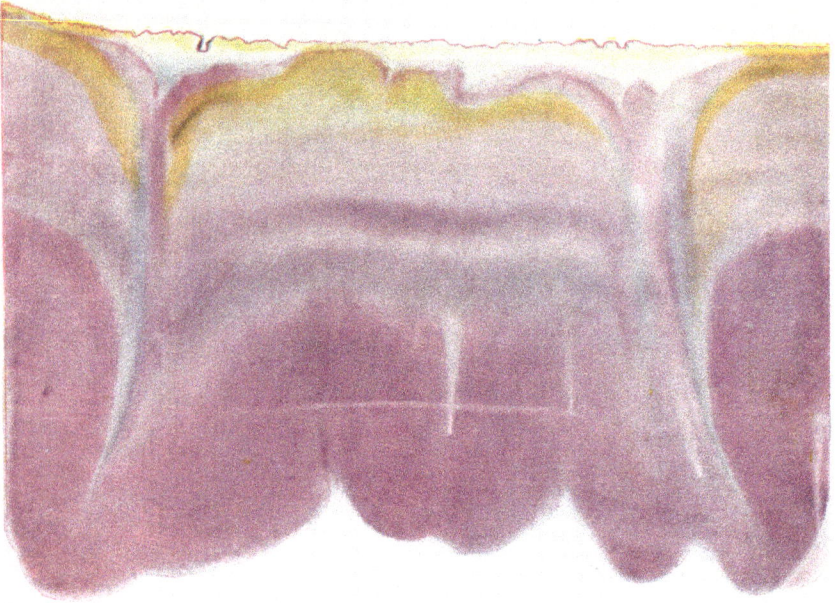

GESTALTUNGSKRÄFTE IN DER STOFFESWELT

GOLDCHLORID

OSTERSONNTAG (MITTAG)

Fig. 30: Gold, Easter. © Kolisko Archive, London.
From: "Spirit in Matter", offprint

Herbert Hahn wrote about this in his book "Von den Quellkräften der Seele":

> [...] One example, which relates to the same time of year, may be cited from scientific observations. On Rudolf Steiner's advice, Frau Lilly Kolisko carried out certain experiments in order to visualise the effect of cosmic forces in the earthly realm. The experiments were carried out on several metals - for example silver, which is connected to the lunar forces. Thus, lead is connected to the Saturnian forces, iron to Mars, and gold to the Sun.
>
> A very high potency of gold was used. A strip of capillary paper was placed in the container containing this liquid and the liquid began to rise in the paper. After drying, different shapes and colours formed on the paper, depending on the type and time of the experiment. Attempts were made to determine whether the respective position of the Sun had an influence on the metal gold, and the samples showed that the Sun's influence on the gold was unmistakable - even when the Sun was obscured by clouds, even when it was night, and even when the experiments were carried out in a dark cellar. Thus, here in Stuttgart, purely from the gold reactions, a solar eclipse not visible at all over this area could be noticed. The individual months were also reflected in the nature of the experiments. The experiments with gold were particularly interesting during the Christmas season, during the 12 Holy Nights. On 23 December, a fairly ordinary gold image was obtained. But on the night of 24 to 25 December, the image began to glow. During the subsequent Holy Nights, the golden pictures are as varied as they are from month to month throughout the year. Even when the fog is oppressive or on rainy days, the pictures of the sun emerge in great purity and beauty during this holy season [...].[218]

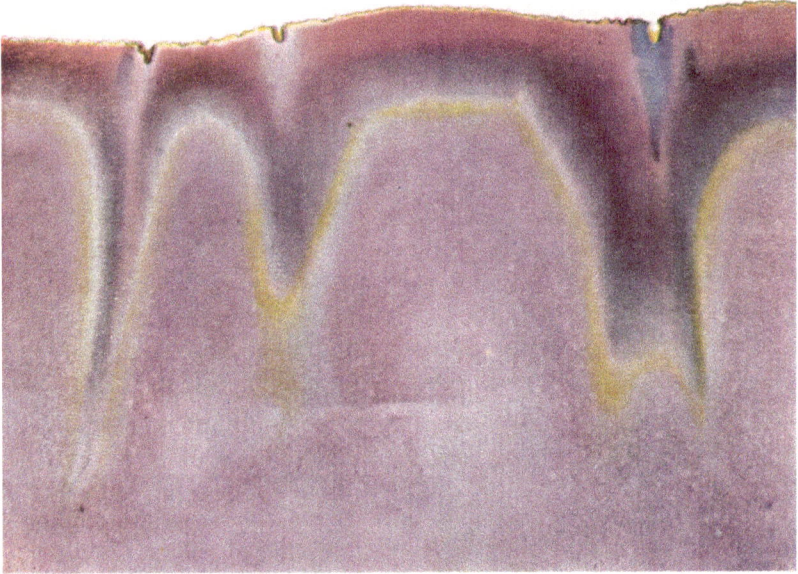

GESTALTUNGSKRÄFTE IN DER STOFFESWELT

GOLDCHLORID

WEIHNACHTEN (25. XII. MITTERNACHT)

Fig. 31: Gold, Christmas. © Kolisko Archive, London.
From: "Spirit in Matter", offprint

14.

Public lectures and scientific controversies

The results of the reported experiments are so far-reaching that the repetition and re-examination of the experiments appears urgently necessary from a scientific point of view. If they are confirmed and no source of error is uncovered, the consequences for the whole of scientific research are incalculable. If inorganic substances really are exposed to the influence of stars, then it stands to reason that the same can be said of living cells. [219]

Frau Kolisko's lecture on the working of the stars in earthly substances with pictures obtained during the last solar and lunar eclipses was one of the highlights of the conference, also in terms of the number of listeners [...]. [220]

Lectures in England and Vienna 1927 /28

The "International Summer School" in England, which Rudolf Steiner had previously attended, were continued by Daniel Dunlop in Gareloch, Scotland, in 1927. From 25 July to 5 August 1927, there were "lectures on health and illness and an extension of the art of healing according to spiritual scientific knowledge". [221] With other members of the Medical Section, Lilly Kolisko was invited to give a contribution about her research.

Several of her publications appeared in English in advance. In a letter to Ita Wegman on l0 June 1927 she wrote, among other things: "[...] Mr Geuter has finished the translation of the smallest entities (1st book) and will discuss it with me tonight. He wanted to go to Obersdorf on Wednesday. But if you wish, he is also prepared to stay here and begin the second book [...]." [222]

The work on the connection between lead and Saturn during the conjunction of the sun and Saturn on 21 November 1926 was translated into English under the title: "Workings of the Stars in Earthly

Substances".[223] Lilly Kolisko's research thus became increasingly well known in England.

Before the International Summer School, Eugen Kolisko was asked to give a lecture on "the efficacy of the smallest entities" at a homeopathy congress in London in mid-July 1927. He spoke about the experimental work at the Biological Institute at the Goetheanum. "The lecture was met with great interest, and Dr Wheeler, as chairman, emphasised afterwards in a very amiable and understanding manner that this provided completely new insights into the mode of action of potentiation, and that in future a much more differentiated application of the individual potencies for the various substances would be necessary in therapy. We were actually facing a completely new field."[224]

In her contribution in Gareloch, Lilly Kolisko focussed on the changes in metal reactions in different constellations.

> "[...] Today Frau Kolisko spoke in the morning and again in the evening, in very good English and very impressively," Margarete Bockholt informed Hilma Walter from Gareloch.[225]

The "Summer School" was an important event for anthroposophical work and the movement in England; it enabled the exchange of experiences and the maintenance of international relations. And so, at the end of the Summer School, a large "world conference" was planned for the following summer (1928) in London, at which many anthroposophical initiatives in different areas of life were to present themselves.

*

At the beginning of 1928, Lilly Kolisko gave a lecture at the Vienna Society of Engineers and Architects entitled "The mysterious forces of nature, workings of the stars in earthly substancies". Otto Myrbach, an observer at the Central Institute for Meteorology in Vienna, reported in the "Neues Wiener Journal" on 7 February 1928:

> A few days ago, large posters attracted the Viennese population to a lecture given by Frau Elisabeth Kolisko at the Association of Engineers and Architects. "Working of the stars in earthly substances!". This sounds so fantastic to the natural scientist that mistrust must have

prevented most experts from accepting the invitation. Nevertheless, the large hall was filled to capacity, admittedly mainly by women, who seem to follow today's school of thought with more love and interest than the more preoccupied male world. For my part, I followed with curiosity, but was not prepared for the fact that the speaker would give us all an hour of deep devotion [...].[226]

Myrbach outlined various of Lilly Kolisko's research findings. She spoke in the Association of Engineers and Architects about her experiments using slides, in particular about the planetary effects in various metal salts, which were detectable during different constellations - the Sun-Saturn conjunction in November 1926 and the solar eclipse in June 1927. Myrbach concluded with the words:

> The results of the reported experiments are so far-reaching that the repetition and re-examination of the experiments appears urgently necessary from a scientific point of view. If they are confirmed and no source of error is uncovered, the consequences for the whole of scientific research are incalculable. If inorganic substances really are exposed to the influence of stars, then it stands to reason that the same can be said of living cells. And for an organism that harbours dissolved metals, the conclusion would immediately follow that the function of such solutions would also be disturbed in the living body by certain celestial positions. [227]

Once again, the term "from a scientific point of view" was used in this article - apparently to express the fact that Lilly Kolisko's work in an anthroposophical institute was not recognised as being of sufficient quality, and very probably not the lecturer herself, since she did not have an academic degree. This was followed by the "World Conference on Spiritual Science and its Practical Applications for the Well-Being of Humanity" in July 1928, during which Lilly Kolisko was back in London.[228]

The biologist Hermann Poppelbaum reported in the German magazine "Anthroposophie":

> This programme [of the London World Conference] provided for such an abundance of lectures, discussions,

presentations, concerts and exhibitions that the out-of-town visitors hardly had time to visit London. Only a sample of this abundance should be emphasised, and this is not to say anything about the value of the others ... Each day was dedicated to a particular goal. The medical day and the day dedicated to new natural research gave the most cohesive impression. On the former, the demonstration of therapeutic eurythmy had a decisive effect; on the natural scientist's day, the beautiful connection between the more fundamental lectures (Dr Wachsmuth, Dr Vreede) and the reports on the researchers' experiments (E. Pfeiffer, L. Kolisko) was impressive. Frau Kolisko's lecture on the working of the stars on earthly substances with pictures obtained during the last solar and lunar eclipses was one of the highlights of the conference, also in terms of the number of listeners [...].[229]

Scientific controversies about "smallest entities"

Lilly Kolisko reported on discussions about her work on the "smallest entities" to Guenther Wachsmuth in Dornach in August 1928:

Dear Dr Wachsmuth! I am very grateful to you for sending me the letter from Dr Schmid-Curtius and would like to give you a brief account of the matter for your information.

Dr Fenner is the editor of the journal "Biologische Heilkunde". Years ago, he criticised my first paper on the effect of the smallest entities in this journal very badly and maliciously. I never responded because I consider it pointless to react in any way to malicious criticism. Dr Schmid-Curtius came across Dr Fenner again by chance some time ago and tried to persuade him but was of course unsuccessful. The gentleman replied that he had checked my experiments with the help of probability calculations and had concluded that the maxima and minima of the curves were still within the margin of error. He had also repeated the experiments and had also come to different results from mine. My work was completely unscientific. He, Fenner, now

Wednesday, July 25

Art

10.15 A.M. LARGE HALL. Anthroposophy and the Artist. BARON A. ROSENKRANTZ.

12 NOON. ART GALLERY.
Exhibitions of Goetheanum Paintings, by various Artists.

2.30 P.M. LARGE HALL. The Artistic Use of Concrete. M. WHEELER, M.A., F.R.I.B.A.

4 P.M. Exhibition of Paintings. MISS HILMA AF KLINT (Upsala). With an address.

5 P.M. LARGE HALL. Music in the East and West. MR. ZAGWYN. Illustrated on the piano by MR. JAN VAN DEN BERGH.

8.15 P.M. LARGE HALL. The Evolution of Music in the light of Anthroposophy (with musical examples). MISS JEANNE DE MARE.

8.15 P.M. RUDOLF STEINER HALL.
Demonstration of Eurhythmy by Artistes of the Goetheanum School of Eurhythmy, Dornach.

Thursday, July 26

Natural Science

10.15 A.M. LARGE HALL. The Earth as an Organism. DR. GÜNTHER WACHSMUTH, Secretary of the General Anthroposophical Society, Leader of the Science Section of the Goetheanum, Dornach.

12 NOON. LARGE HALL. Spiritual Aspects of Astronomy. FRL. DR. E. VREEDE, Leader of the Astronomical and Mathematical Section of the Goetheanum, Dornach.

3 P.M. LARGE HALL. Experimental Researches into the Formative Forces of Nature (illustrated by lantern slides). EHRENFRIED PFEIFFER. (In German. English translation by G. Kaufmann, M.A.).

5.30 P.M. RUDOLF STEINER HALL. Demonstration by the English students of the Rudolf Steiner School of Eurhythmy, London.

8.15 P.M. LARGE HALL. Influences of the Stars on Earthly Substances (Scientific experiments at the Biological Institute of the Goetheanum). Illustrated by Lantern Slides. L. KOLISKO.

12

Fig. 32: From the programme of the World Conference in London (25/26.7.1928).
© *Ita Wegman Archive.*

wanted to present his material against me at the Nature Researchers' Conference.

Dr Schmid-Curtius then visited me in Stuttgart and said that he would personally attach great importance to Fenner being proven wrong with his claims. Could I give him some information for this purpose? Dr Schmid-C. mentioned that this matter was his *bread and butter.* His entire work was based on my work and stood and fell with it [...].

I have now finally shrugged off Dr Schmid-Curtius and enclose a copy of my letter to him. The *Junker* matter, on the other hand, is very interesting. But it would take too long if I were to report on it in writing. Perhaps I can tell you more about it orally. For today, I would just like to quote a few passages from his latest publication, he sent me himself a few weeks ago.

"Die Wirkung extremer Potenzverdünnungen auf Organismen" by Hermann Junker with 21 text illustrations ... (Offprint from: Pflügers Archiv für die gesamte Physiologie des Menschen und der Tiere, 219. Vol,

5th/6th issue) Page 660:

"Already in my first communication I reported briefly about Kolisko's experiments in 1925. She also investigated the effect of extremely diluted substances (up to 1:10/30) on germinating wheat. To say it straight away: Kolisko and I came to the same conclusions, independently of each other, on different objects and with different substances. She has recently published a second work on the same topic. Since Kolisko's publications will not be easily accessible to the general public, it is probably worth briefly describing her experiments. First a few words about her earlier communication ... the experimental arrangement and Kolisko's way of working appear to be extraordinarily exact, as the results obtained prove ...

Junker then described the experiments from the second publication:

"The reason for these experiments - which, incidentally, also appear to be technically very precise - and the evaluation of the results arise purely from considerations and ideas. She blindly ignores the extraordinarily interesting and valuable details of her curves and merely focusses her attention on the appearance of the entire curve image. In doing so, she arrives at ideas and meanings of a 'spiritual-scientific' nature about it seems pointless to lose words. Even cosmic connections and the like are sought and found. These are, of course, speculations that are completely absurd. It is a great pity that she does not judge her beautiful results with a more sober eye!"

Another passage taken from the "Summary":

The purely factual results of *Kolisko*, who carried out the same investigations on wheat germ, were in complete agreement with the findings on pharmacopeia.

It could be shown (also based on Kolisko's experiments) that two or more such series can be brought together ("grafted on top of each other"). exert their effects independently of each other. The result of this is a summation of the effects of the individual dilutions.

Seybold's criticism of the findings of *Krakow, Kolisko* and myself, <u>could not be completely rejected as unfounded.</u>

You will probably realise from the few random samples that Junker has largely made up his mind. He can't help but defend me against Fenner's attacks. For Junker, too, it means a "Renommé" question. Because if Fenner's opinion prevails that all minima and maxima below the error limit are to be seen with probability calculation, then this would apply to a greater extent to Junker, who worked with much less material than I did and whose deflections are also much smaller than my curves show.

I really ask: Why should Schmid-Curtius appear? I think Mr Junker and Mr Fenner will argue much better together. I would really enjoy listening to them, but that will probably not be possible. Yours sincerely, L. Kolisko [230]

Lilly Kolisko's reply of 8 August 1928 to Schmid-Curtius was also preserved:

Honourable Dr Schmid-Curtius

I would like to come back to our last conversation in

Marienstein regarding the natural science conference in Hamburg. After careful consideration, I have come to no other conclusion than the one I have always expressed to you: I do not think it is appropriate for you to advocate potency work at the natural scientists meeting. Nothing will come of it. You think that my latest experiments, carried out in the open air, must be absolutely convincing for the people there. I believe that anyone who has already not given credence to the two books that form the basis for the latest work have no reason to give credence in the new trials in the open air.

They also wanted me to ask one of our mathematicians whether probability theory could be applied to biological experiments at all. Unfortunately, it was quite impossible to speak to one of the gentlemen in the last few days of July (just before the end of school), and now during the holidays there is certainly no opportunity to do so. This question will probably not be decided any time soon.

In Marienstein you also enquired about Dr Junker, who had been in correspondence with me years ago about his own experiments with potencies. Coincidentally, I have just now received a paper from him on the same subject. He has continued his experiments, again only up to the 27TH potency. He got curves that are similar to mine. In his last paper he discusses my experiments in detail and completely recognises their accuracy. He is only dissatisfied with the part of my paper that deals with anthroposophical views. Indeed, in the last part he even defends me *nolens volens* against Prof Seybold's attacks and says that his objections against me and Krakow, and therefore also against (Junker) himself, are completely unfounded.

Junker is in Hamburg, most likely attending the naturalists' meeting, and I think it safe to leave it to Junker to argue with Fenner. This has far more chance of success and works much better if the defence comes from the outside world than if it comes from within own camp.

As far as your request for 300 selected grains of wheat

is concerned, I will gladly fulfil it, but you will have to be patient until I receive this year's wheat. The harvest is just around the corner and I have just written to Mr Stegemann to ask him to give me some of his organic wheat and some artificially fertilized wheat. Perhaps it would be interesting if you could also carry out trials with these two varieties. Yours sincerely, L. Kolisko [231]

At the same time, in August 1928, a homeopathy conference was held in Stuttgart, at which Professor Hans Much from Hamburg was one of the speakers. He mentioned experiments that were strongly reminiscent of those carried out at the "Biological Institute", but did not provide any sources; a presentation of his contribution appeared in the "Stuttgarter Tagblatt". Lilly Kolisko wrote about this in a letter to Ita Wegman and Guenther Wachsmuth:

> [...] I spoke to Dr Schickler about this, and he is of the opinion that something should be done. He would be happy to write to Much saying that it was very astonishing that he had given a lecture here in Stuttgart and reported on experiments that had been carried out for years at the Biological Institute at the suggestion of Dr Steiner and about which 2 publications had already appeared, and so on.
>
> I would now like to ask whether you think it would be desirable for something to be done about this matter. It could be that Professor Much becomes unpleasant when he is annoyed. On the other hand, if you are of the opinion that Much should be expressly made aware of the work (which he certainly knows), then it should be done immediately. So it would be good if I could receive your opinion soon.[232]

No reply from Ita Wegman or Guenther Wachsmuth has been preserved in the estates.

15.

Autumn 1928 and personal situation

[...] Then there's one more thing I'd like to tell you. Your wife was quite unwell when she was last here [...].[233]

Lilly Kolisko was overwhelmed in the autumn of 1928 with the abundance of her tasks - research, teaching at the Waldorf School, continuing work on publications, Class lessons, family and on top of that the serious conflicts in the Anthroposophical Society. She was in poor health and was invited to travel to Gnadenwald and Italy, from where she wrote to Ita Wegman, also about her state of health. She had back pain and pain in the spleen area and suffered from fevers.

On 5 September 1928, Margarete Bockholt informed Eugen Kolisko:

"[...] Then there is one more thing I should tell you. Your wife was quite unwell when she was last here. We would really like her to take a longer break and stay here with us so that we can take good care of her. Your wife was actually not averse to staying here for a longer period after Michaelmas, as she herself feels that it is necessary. It is very important to us that this happens, and Frau Doctor [Dr Wegman] also wants to write to your wife about this. But we would also like to ask you to influence your wife in this direction [...]."[234]

Ita Wegman actually wrote to Lilly Kolisko three days later along these lines:

Dear Frau Kolisko!
[...] I would like to reiterate my urgent invitation to you to stay here with us after Michaelmas so that we can take a thorough cure. I believe that it is probably best here and I am also convinced that a great deal of fruitful work can come out of it. I have just spent a few

days in Gnadenwald and, as you wrote, had a very good impression. Well, goodbye, dear Frau Dr Kolisko, until Michaelmas, get plenty of rest until then and in the meantime continue to take your medication as we have discussed.[235]

In response, Ita Wegman received a long, very personal letter that Lilly Kolisko had written on 11 September 1928, immediately after receiving Wegman's letter:

> Esteemed, dear Dr Wegman!
> This morning I received your friendly letter, which I will try to answer right away. I have wanted to write to you several times, but something always stops me. Sometimes things can't be written down but can only be spoken. During my journey with Frau. Leroi, I have always thought of you and written down many things, a little bit each day that I would have liked to tell you. But it is very difficult for me to hand over what I have written to you. I have had to think again and again of your words: We want to be completely honest, at least in private, and tell how things really are. I was shocked by how quickly and confidently you diagnosed my organs as defenceless and that every event strikes directly into them because there is no protective armour around them. But now it has brought me to a question of conscience: shouldn't I lay my soul before you just as openly as my body? Isn't it necessary, in order to heal the body, to also reveal the damage to the soul? I know for sure that you will answer me: It's a question of trust.
> Dear Frau Doctor, I have examined myself and know that I would never have come to you and never spoken to you, as I did last time, if I hadn't had trust and something that is far more: a great personal connection and love.
> In your letter, you ask how my health is now. It changes. I felt a bit better on Friday and Saturday and thought I might be able to overcome it again. On Sunday it got worse again, so that I woke up in the morning with the burning pain in my back that hurts my spleen and liver. Sometimes it gets so intense with stabs that seem to pierce the body, that when I'm walking, I have to stop for a while and then only walk very slowly and in constant pain.

But last week something very strange happened to me, I wanted to write to you straight away, but then I thought I was taking up too much of your time with personal matters. But it is interesting. It was around 3 o'clock in the afternoon. I was incredibly tired again, went back to my room upstairs in the Biological Institute, lay down and did an exercise that Dr Steiner had once given me. I must have fallen asleep for a short time and woke up again. In other words, I was already fully conscious but could not take hold of my physical body. I tried to sit up - I didn't succeed, I tried again but kept sinking back down. I was able to open my eyes halfway and looked through the window opposite me. Again, I realised that I was unable to open my eyes fully, there was a heavy pressure on me and my eyes seemed swollen. It was a strange state. I inwardly said the verse: S.[ervus] Michaeli[s] s.[um] p.[er] m.[eam] v.[oluntatem] [I am Michael's servant by my own will.] to relieve the tension - in vain. I straightened up again and managed to sit half upright when an uncanny force hit my back. I clearly felt as if two hands were reaching through from behind and gripping my organs. It was no illusion. I can tell you exactly where the hands were gripping me.

They compressed the internal organs, liver, spleen, and then pushed upwards and seized my heart. A raging pain ran through me. Unfortunately, there are no words to describe what happened. I was being pulled backwards with great force and I had no strength to resist this elemental force. I fell back and began to pray the Rosicrucian verse. My heart stood still, and now something quite unbelievable happened to me: I could not remember a part of the Rosicrucian verse. I could only think of two. I could find all the corresponding movements, but not the words. So I repeated one part and then said the second. While this was going on - it must have taken only a few minutes - I heard Sister Ruth coming up the stairs. I was frightened because if she had called me, I wouldn't have been able to answer and now I visualised the situation if she opened the door and found me lying there. I was ready to send for a doctor at that moment, possibly my husband. I got up gingerly after about 10 minutes. Fortunately, Nurse Ruth had only gone into the theatre room to do something.

Staggering, I went to the table and tried once again to grasp my situation. I was still experiencing severe pain, as if my ribs had been broken, and now I tried to say the Rosicrucian verse again, and again I couldn't remember the third part. It is strange which spell I could not find. I spoke I. Ch. M. twice, then P. SP.S. R. And I could not find E. D. N. I would like to emphasise once again that any deception was impossible. It was a thoroughly alert experience. After half an hour I went into town because it was necessary, met Miss Uhland from the Waldorf School, we exchanged a few phrases, I clearly felt that I had not yet completely taken possession of my body. Frau Uhland suddenly said: "If only I could I'd put you to bed straight away. You have snow-white lips and deep circles around your eyes." So an unbiased person could see this objectively.

It seemed remarkable to me that I had just forgotten the first part of the saying, secondly that there are demonic powers that want to prevent me from immersing into my body. This shock still has an effect today. I have become clumsy. Objects easily slip out of my hands, which didn't happen before. Even very light ones: I now have to bend down all the time to pick up pencils, thermometers, keys, etc. I am seriously wondering what to do if this happens again.

I am very pleased that you have been to Gnadenwald. You will certainly have found everything I wrote about it confirmed.

I look forward to the opening of the [second Goetheanum] building with trepidation. I keep thinking that there will still be major catastrophes. I believe that many things will break before the building is opened. It is a great mystery to me that it was possible for a Waldorf pupil to a Waldorf pupil to commit suicide. What does that mean? What consequences will it bring for the school? They will do everything they can to prevent it from becoming public. Tomorrow is the memorial service at the school. Who will protect us now from this case bringing other children into its wake? We have so many unstable elements.

Dr Kempter is resigning from his position as branch chairman of the free Society. He was with Schlichter in Schliersee and came back transformed. He finds that anthroposophy is ossifying into doctrine and cries out for the living spirit. It is possible that Dr Kempter's resignation will also result in Mr Ege's, in which case Dr Lehrs will be alone again. Events will regress. But they can't stop there. And will people find what is now required of them?

Now I come to the last part of your dear letter. You asked me to stay in Dornach after Michaelmas for a cure and to work with you. Oh, dear Frau Doctor, how long I have longed to be allowed to really work with you. If I were to follow my heart, I could only happily agree. Yes, I even believe that it is an objective necessity, that it is a spiritual requirement, the intimate co-operation with you.

Against this is that: I am teaching in the school, there is now one who can take over.

It is not good for me to interrupt the work at the Biological Institute. Continuing part of it in Dornach would involve additional expenditure and we would be left with nothing again. The money from England has already been used up to pay debts, heating material for the winter and filter paper. The agricultural work makes it necessary for me to make the preparations myself now in the autumn, otherwise I won't be able to do any experiments in the spring.

Geni cannot be left alone for so long. A carer would have to be found for her at least for the afternoons. And finally, my own stay in Arlesheim will not be financially feasible at the moment.

I can't find my way through all these difficulties; perhaps there is still a way out. In any case, thank you from the bottom of my heart for your kind invitation. For my part, I will try everything, because it would be nice and good to work with you, even if only for a short time.

With   warm   greetings, with grateful respect,
L. Kolisko[236]

Ita Wegman, however, obviously did not give in, emphasised the need for real recovery and invited Lilly Kolisko. Ten days later, on 1 September 1928, she wrote again:

> Esteemed, dear Frau Doctor, I have just received your letter, which I have been waiting for each day. Firstly, please let me say that I feel very ashamed. It was not good of me to put money alongside the spirit.
>
> But I really want to do everything I can to overcome myself. Yesterday I wrote to Dr. Bockholt that I would like to come a little earlier, perhaps as early as Wednesday evening. I would like to continue my experiments in Arlesheim, and I feel a strong urge to come a little earlier to talk to you. If that is okay with you, I would stay until about 15 October, then go to Stuttgart for a few days to complete my autumn work, and then, if you think it's right, come back to you. I haven't found anyone for Geni yet, but I'm constantly looking [...].[237]

Fig. 33: The Second Goetheanum in Dornach.
© Documentation at the Goetheanum

16.

The opening of the Second Goetheanum

> However, as I was now directly asked to contribute to the opening ceremony with a lecture, I naturally expressed my willingness to do so. I wrote that I would like to give a slide lecture on the reflection of the festive seasons in scientific experiments, as this seemed to me to be the most valuable that I have worked on to date.[238]

I n Michaelmas 1928 the Second Goetheanum was opened in Dornach with a large conference in which Lilly Kolisko took part.

Before she could travel to Arlesheim, she made intensive preparations for her slide presentations. She chose the topic: "Reflection of the festivals in scientific experiments". As she had made capillary dynamolysis pictures with metal salts every day for years, she was able to observe that the pictures correspond to the annual cycle and that they are different during the Christian festivals than during the rest of the year. Now she wanted to speak about this at the opening of the Second Goetheanum; at Easter 1923, three months after the catastrophic fire, Steiner had begun his lectures on the spiritual breathing of the earth and the festive seasons - and had been preparing the new building on his sickbed until the very end.

Lilly Kolisko wrote to Guenther Wachsmuth in advance:

> Dear Mr Wachsmuth!
> A few days ago I received the conference ticket, for which I thank you very much. Two days later I received a supplementary ticket for a mystery drama that I would have missed due to my lecture. I must confess that I am very touched by your care! With the workload that is surely resting on you now, you still find time to think about whether some of the speakers are perhaps missing out a little because of their presentation!

I realised to my horror today that when I mentioned the topic I forgot to add that it was a slide lecture. Frankly, I took it for granted, and it only occurred to me now that it might not be so obvious to you. However, there will only be nine times 12 slides, so there won't be any problems with the equipment. But can projections also be used in the new Goetheanum, or only in the Carpentry workshop? Yours sincerely, L. Kolisko [239]

Her letter was forwarded to the engineer Paul Eugen Schiller, a co-worker of Ehrenfried Pfeiffer at the Dornach Glass House, for his attention "with request for further instructions". Decades later, Lilly Kolisko wrote in her memoirs about the difficult production of colour slides on glass plates at that time - and about the complicated conditions in Dornach:

> After the [controversial] correspondence between Dr Vreede and Herr Steffen, the speaker issue during the Michaelmas conference was settled in such a way that an invitation was received from Dornach to indicate a topic for a lecture. I also received such a telegram. I must admit that I would never have responded to the invitation printed in the newsletter. It would have seemed immodest. However, as I was now directly asked to contribute to the opening ceremony with a lecture, I naturally expressed my willingness. I wrote that I would like to give a slide lecture reflecting on the festive seasons in scientific experiments. This seemed to me to be the most valuable of what I have worked on to date. It would not be easy, as the time available to me was very short to make the necessary preparations, but I would do everything I could. To be able to give this lecture a large number of colour slides had to be produced. In 1928 it was not yet possible to produce colour films.
>
> My assistant first had to produce the normal negatives, then attenuate them so that only a faint outline was visible on the glass plate, and then Miss Raisch had the task of painting the glass plates based on the original experiments. It often took several hours to create a single image, and about 80 exposures were needed for the lecture.
>
> Work was carried out day and night at the Biological

Institute at the Goetheanum in Stuttgart to complete it. For a lecture like this, a great deal depends on the projection, especially on a good device. To be absolutely sure, Mr. Kaiser, who was responsible for the photographic work, went along to perform the projection himself.

The lecture had to be held twice, as the number of members who had come to Dornach made it necessary to hold the lectures twice. Once the lecture was to take place in the Carpentry workshop and once in the Goetheanum. The first lecture took place in the Carpentry workshop, and it can be said that the projection was quite successful. However, Mr Kaiser was of the opinion that the lecture in the large hall of the Goetheanum would not be possible with the projector equipment provided. Only the front rows would see reasonably good pictures, further back you would only see blurred pictures. It was essential to find a better device. He contacted the member Mr Paul Eugen Schiller, who was responsible for that. The answer was that there was no better machine. You couldn't get a better machine in the whole of Basel. You couldn't borrow one either. We were desperate that all the hard work would be in vain and that we would have to give a bad lecture at the opening. I asked Mr Kaiser to go to Basel and try to find a better machine somewhere. I would cover the costs. If there was no machine in Basel, he could go to Zurich.

Mr Kaiser returned from Basel a few hours later and reported that it was of course possible to hire a good machine. He had found 4 machines and would now go to Mr Schiller and let him know. Well, he was again unsuccessful. The answer was that the necessary electrical connections were not available. Not satisfied with this, Mr Kaiser went backstage and found that all the necessary connections were there. He asked his friend, because the person in question was his friend from Stuttgart, where they had worked together: "Why don't you want the lecture to be projected well? I refuse do a bad projection when a good one is possible. Why don't you want it?" The answer was: "Frau Kolisko already has a good reputation, she doesn't need to improve it!" It sounds almost unbelievable, but it's true. Mr Kaiser refused to project. Mr Schiller, who prevented me from using the good projection equipment, had to take over the projection himself during my lecture. I'm including this experience here because it's a good example of the resistance I constantly had to deal with. It can certainly also be categorised as "inner opposition" ...[240]

In November 1928, Hermann Poppelbaum published an article in the journal "Die Drei" entitled "Zur Methodik der Kolisko'schen Phänomene"[241], in which he discussed Lilly Kolisko's work on potentiation and capillary dynamolysis. At the end he wrote about the experimental results she presented at the opening of the new Goetheanum:

Frau Kolisko has now even been able to demonstrate how even the annual festivals (Christmas, Easter, St. John's Day, Michaelmas) are clearly revealed in the inner processes of metal solutions. When these new experiments, which the researcher presented at the opening of the new Goetheanum, are published the *human* significance of the new experimental methodology will emerge beyond all doubt. Kolisko's rising pictures pose the crucial question to every natural scientist and every person in general: whether, while maintaining critical awareness, they want to call upon their own creative capacity in the face of these pictures, or whether they want to remain rigid and unchanging before these works of art. The world itself paints, with moving vividness, on the modest white strips of paper, the first stage of the revelation of the supernatural: the lively, colourful, varied, yet expressively eloquent image. *We are faced with the significant event that nature itself, permeated by spirit, calls upon man, who offered it the canvas, to attain the power of imagination.*

The methodology of Kolisko's phenomena leads to the beginning of the path to supersensible realisation - without any arbitrary action on the part of its discoverer, merely out of spiritual consistency. It is a hint from the spiritual world itself. One can hardly overlook it.[242]

*

On 28. October 1928, just under four weeks after the opening of the new building, Ita Wegman wrote another long letter to Lilly Kolisko, in which she asked, among other things, for a further contribution to the journal "Natura". The disputes surrounding the Anthroposophical Society in Germany and Dornach[243] were also expressed in it:

> Dear Frau Kolisko, I would like to know how you are doing. I'm worried that you haven't written to me, which I must take as bad sign. But of course, it could also be that you have so much to do and that you are so absorbed by the events in Stuttgart that you don't get round to writing. But before I leave for Wroclaw tomorrow, Monday, I would like to have written you this letter.
>
> I will be travelling from Wroclaw to Jena and staying away for a week.
>
> We have a tough time here. Miss Vreede is not having an easy time. Next Wednesday, Boos and Dr Grosheintz will report on what happened the General Secretary's meeting. It's completely crazy. What we wanted to avoid [should have been avoided] in a general meeting is now being communicated to the members in a nuanced way, so that it is even worse than before a General Meeting. First through Unger's letter, then through Boos. I don't know what else could happen on Wednesday. I'm departed and poor Dr Vreede is all alone. We're the bad guys, of course. Fortunately, yesterday we received a letter of defence for Miss Vreede from Hohlenberg, which is intended for the General Secretaries. Steffen is now being attacked again. It will go on like this; a group don't want peace.
>
> What I would like to write to you now, however, is to ask you write something about potencies for Natura. This is so important that it should be published as soon as possible. We can then create a lot of publicity for this magazine, which is entirely at your disposal. It is also important that doctors are well informed about these things, because our medicines are based on these potentiation's. And you would be doing us a great

favour, because we would like to have your work published in "Natura". We are not flooded with contributions and are doubly pleased when they are willingly shared.

I can tell you the following about England, and perhaps you should also tell your husband that. Strange things have also happened there. A National Executive Council of 12 members was formed, without a general secretary. Dunlop had a way of telling Collison the truth and calling everything they did a lie, and then he left the room thus excluding himself. Acting thus he actually made the re-election of C[ollison] as president impossible. I don't think the situation as it is now in England is bad. It is certainly clean and opens up new possibilities.

More than half of the twelve-member Council are reasonable people. They won't be able to be active because they will always inhibit each other. But I think Dunlop has gained a lot of freedom mainly in the medical field. Wheeler has remained treasurer as the only official functionary of the National Council. Kaufmann is a translator, but it is generally assumed that he will become secretary of this Council because he has many friends there, so that when it comes to a vote, he will get the most votes. There must have been dramatic scenes, at least from the accounts I received.

But I very much hope that you are doing reasonably well and that I will hear from you soon. Please accept my warmest regards, and please also give my best wishes to your husband and Geni.[244]

On 6 November 1928, Lilly Kolisko replied:
Esteemed, Dear Dr Wegman!
You will probably be very surprised that you haven't received a letter from me yet. However, things have not been going well for me at all. When I got to Stuttgart with Mr Kirchner, there was just about time to see Dr Rittelmeyer. I was very interested to hear what he had to say about all the Society matters. I survived the car journey quite well and so I went straight to see Dr Rittelmeyer. I must admit that I was terribly depressed by the way he spoke. My liver was constantly riddled

with very fine but penetrating stitches [...]. From then on, I felt increasingly worse, especially mentally. I was overcome by tremendous depression and apathy. It seemed as if I was simply overwhelmed by the spiritual atmosphere of Stuttgart. I can't describe it, but it was terrible. Physically, I wasn't so bad—but mentally. I couldn't muster up any resolve. Everything I had planned to do in Dornach was as if blown away! So I was glad that, since the Herr Kaiser was still on leave, I had to work outside. I dug the potatoes out of the ground, made the agricultural preparations, etc., since I couldn't do any mental work at all. Now this hard physical labour was too much for me again [...].

Last week I gave a small group (about 30 people) of technicians, chemists, physicists etc., invited by the machine manufacturer Kleemann, Obertürkheim, a lecture with slides on the "influence of the stars", followed by a discussion. This lasted until 12 h at night. It was very interesting. Dr Maag was also there!

Then I changed the silver book to such an extent that the publisher could calculate its cost. Now I'm starting the potency work, but I don't know how quickly it will go.

In the meantime, my husband has also collapsed again. These endless meetings and conferences are wearing him down completely. Plus the lessons, school doctor's duties, university courses, further education school, teachers' seminar, private conversations - it became too much. He stayed at home for "a couple of days" and hoped that he would now be able to carry on again.

It's bleak here. I hope you are feeling better! Best regards, yours grateful L. Kolisko [245]

Eugen Kolisko's health deteriorated. As early as June 1928 - after an accident in Hamburg - Ita Wegman had recommended that he come to Gnadenwald or Arlesheim to recuperate. He finally went to Gnadenwald in the late autumn of 1928 (near Hall by Innsbruck, Tyrol), where Ita Wegman ran sanatoriums, the medical management of which she entrusted to the doctor Norbert Glas (from the circle of the "young physician").

On 24 November 1928, Lilly Kolisko wrote to Ita Wegman after a visit to Dornach:

Esteemed Frau Doctor!

When I got home in Stuttgart last night, I found a letter from my husband in which he reported that the vets had first diagnosed swine fever in the pigs, then something else, and finally they said the pigs had worms and sent a worm remedy! He treated with Distempo, and apart from two pigs and a piglet, none of the animals died. So everything seems to be happily over.

In my last letter, I had written to him that I would like to visit him for a day. I received a very cheerful reply. He had already sent me all the train connections and fares. He also told me that there was a second bed in his room so that I could sleep there. Then he asked if I could come straight after my lecture in Dornach. So it seems that he would like to see my visit.

Unfortunately, Geni wasn't in a good mood. Miss Martha told me that she couldn't go to sleep at all, that she woke up at night and always had to keep the light on in her room because she'd had such bad dreams about me. It is indeed very strange what Geni dreamt: I had stayed out too long, and in her dream she asked Martha why I hadn't come back yet. Martha replied, "I would be hanged in the Carpentry workshop". Then she woke up and kept crying that she was afraid something would happen to me. I can't explain how a child can dream like that. I hardly dared to tell her that I would have to go to Dornach again and then perhaps to her father in Gnadenwald. She immediately burst into tears again and asked me not to go away for so long.

In the evening, when I put her to bed, she started crying again, saying that she didn't want me to go to the laboratory, that she couldn't sleep at all. When I asked her why she couldn't sleep, she explained that she was always afraid that the laboratory would collapse, that the building was so badly built and had no foundation stone. Again, I find it strange that Geni comes up with the idea of the foundation stone. A long time ago, I once told her that I had learned that the research institute had not received a foundation stone from Dr Steiner. He had promised to lay it, but it never materialised. From time to time, this played a certain role in the conversations of former co-workers at the research institute, namely that the research institute did not receive a foundation stone from the doctor. Where does this fear suddenly come from in the child? I also found her facial expression has changed. Should I take this fear into consideration and not go to Gnadenwald, or what can I do to ensure that the child sleeps at night when I'm away?

I had an extremely unpleasant dream about Mr Steffen tonight, so I woke up in a deep fright and couldn't fall asleep. I hope nothing happens during my lecture in the Carpentry workshop. Yours sincerely, with grateful respect, L. Kolisko [246]

Ita Wegman answered on 26. November 1928:

Dear Frau Kolisko!

I received your letter today and I am in a hurry to reply. I would like to suggest to you that Geni's Christmas holidays be planned now and that you bring her here with you, even to Gnadenwald. This earlier holiday will not do her any harm, mainly because she is sleeping so restlessly now, and it could also be good for her in Gnadenwald. Then I also think it would be nice if she could go for short walks with her father and you, which [you] don't have time for in Stuttgart. It would be a pity if you didn't go through with your plan. Give the girl Martha a holiday too and close the house. I have a very strong feeling that it's the only right thing to do. Please do it! With heartfelt greetings, Dr Wegman [247]

Lilly Kolisko followed Ita Wegman's advice. On 19 December 1928 she wrote from Gnadenwald:

Esteemed, dear Frau Dr Wegman! I'm sure you are very surprised that you still haven't heard from us. But I've been waiting, because unfortunately I don't have enough positive news to report.

Geni feels very comfortable here. She is sleeping well and eating more. Despite all this, she still looks pale and is not putting on weight. Frau Dr Glas is giving her curative eurythmy and Waldon II. So far Geni has not had any meat, although she is very keen on it. Dr Glas suggested meat twice a week. However, I would like to get your advice on this beforehand.

My husband is still extremely unstable. Every little thing throws him off balance. He was upset about the letter from Leinhas, who wrote about Stein in connection with Herr Grone. He couldn't make up his mind to answer the letter or to make the other decision not to answer it.

Then came the circular from Rittelmeyer. Then a letter [from] Stein about the archive. Then there was the Leinhas article in the newsletter with comment[s] about the national economics course. Everything upset him. I had long, long conversations, but he always came back to the original question: I finally advised him to discuss everything with you again in Dornach. We would now have to take the consequences for what had happened in the past.

I drew up a completely new programme for Germany, which gradually met with his approval. I can't write all that but perhaps you will also like it. He likes going to Dornach because he has a deep connection with the Werbeck-therapeutic singing - but at the same time he is afraid of all the people who will ask him questions. He is afraid of having to attend the General Secretary's meeting on 2 January and, on the other hand, does not have the courage to go to Stuttgart on 1 January.

His indecisive state is really a problem for me. He must decide somehow before he returns to Stuttgart. I was planning to go to Stuttgart on 18 December, spend Christmas there – and carry out the experiments over the 13 nights in Stuttgart. My husband wanted to come to Stuttgart on 23 December, go to Arlesheim on 27 December, and return to Stuttgart on 3 January. For my work, it would have been the only possible option. This division, however, left him in a very unsettled state, so I had to make a sacrifice with a heavy heart and break up my research. There is nothing left but for me to stay by his side now and accompany him everywhere until he has found his way out of this unfortunate state.

We have now agreed on the following. We'll spend the Christmas holidays here. On the 27 December my husband, probably also Dr Glas, and I will go to Arlesheim. We'll leave Geni here until 6 January. Heydebrand and Wilke will arrive tomorrow, who can keep an eye on things and take Geni to Stuttgart. Please, if you can, write to me and let me know if you agree. Perhaps I could get a room with my husband, but that would not be absolutely necessary. He shouldn't go to the general secretary's meeting. He might do something foolish. If you advise him against it, he might go to Stuttgart with me on the first.

As far as the work is concerned, I have made every effort to make as much progress as possible on the Potency-work. I hope to have an outline of it ready by Christmas. I can't do it completely because I couldn't get all the material.

My husband dictated a few things to me, like the Michael lecture, but I must copy it all out on the machine before I can say whether it's any good. I had promised myself more, but there was no enthusiasm for it. He complains about his inability to concentrate.

I am writing all of this to you in full confidence and must leave the rest for a verbal conversation. It will take a lot of tact and love to get everything on the right track.

Best regards from your L. Kolisko [248]

The Koliskos' situation remained extremely tense in the midst of the severe political and social crisis at the end of the 1920s - but their work continued. Looking back on the situation of her husband and ultimately her small family, Lilly Kolisko wrote in her biography:

[...] For Dr Kolisko, the events of the Christmas Conference were a living event that always stood before his soul. He had placed his whole life in the stream of the anthroposophical movement. It is fair to say that a "private life" did not exist for him. He worked from early in the morning until late the next day. He taught the many scientific subjects at the Waldorf School and then, after the end of the lessons, hurried over to the doctor's room, which was always overcrowded and often kept him busy until 2 in the afternoon. In the afternoons - when he had free moments - there was his work in the archives with his extensive correspondence, then meetings with the Executive Council of the National Society, as long as it still existed, then private discussions with individual members of the Anthroposophical Society who sought advice from him, in the evenings usually a lecture, whether on Wednesday evenings to the branch members or on other evenings in the university courses. An extensive *welfare activity* had developed over the years: the so-called "Dutch kitchen", holiday trips for children in need of recuperation were organised within Germany, but also to Switzerland, where he was helped a great deal in arranging accommodation for underprivileged children by Dr. *Vreede,* then came the creation of a fund for the doctor's room, which was to enable the free distribution of medicines to the underprivileged. He gave courses at the Eurythmeum for the eurythmists studying there until Frau Marie Steiner dismissed him ... What could not be done during the day had to be done at night. Preparing for lessons always meant working until 2 o'clock in the morning or even later. In addition, he also gave preparatory courses in these subjects to teachers who had difficulties in natural science lessons, which had to take place at 7 o'clock in the morning. He helped some of his colleagues - especially in chemistry lessons - to prepare the experiments for their lessons the next day in the chemistry room after the teachers' conferences, which themselves often lasted until 2 o'clock in the morning.

It sounds almost unbelievable that when he comes home dead tired, his only complaint was: "I've been sleeping too long. I should sleep less" ...[249]

17.

Continuing Research Tasks 1929

[...] The difficulties are always the same. The natural sciences section is rightly responsible for agriculture, the medical section for medicine, and if both areas are covered in a paper, both must be represented [...].[250]

[...] Two rhythms interpenetrate: that of the year, the sun, and that of the month, the moon. We can see the phases of the moon in the greater contouring and fullness of form in the waxing as opposed to the waning moon, somewhat corresponding to the difference between the individual "day pictures" and the "night pictures". However, the effect of the sun during the year is revealed to us in a particularly characteristic way when we study it closely. What arises there in different seasons, changing from month to month, cannot be explained by purely physical causes such as temperature increase and the like, but true formative forces appear in the meaning that spiritual science gives to this word [...].[251]

[...] The fact is, however, that Dr Hauschka has published such experiments in "Natura", which therefore originate from me and have not yet been published by me. In the outside world, there is a certain name for such an offence, which I need not mention to you.[252]

When she returned to Stuttgart at the beginning of 1929, Lilly Kolisko was - in addition to her research work and teaching natural history to the 11th grade at the Waldorf School - writing an article for the so-called "Bauernkalender" (later "Sternkalender") of Elisabeth Vreede's Mathematical-Astronomical Section, organising the publication of potentiation works and completing a major work on silver and the moon.

Her work was cross-sectional. The potentiation work was of great importance for both the Medical and the Natural Science Sections, but her research was particularly important for work in the biodynamic field.

The name of her institute, "Biological Institute at the Goetheanum of the School of Spiritual Science in Dornach", gave an indication that it was not subordinate to a particular section, but was directly affiliated to the School of Spiritual Science.

Her letters provide a clear picture of her thoughts and endeavours in this regard:

> Honourable, Dear Frau Doctor [Wegman]!
> I am enclosing a copy of a letter to Dr Wachsmuth because it says something about the potency work. It was actually finished a fortnight ago, but my husband finds it somewhat polemical and thinks that some passages should be removed. Grunelius also read it and had the same impression. There's no other option but to rework it, as taking out individual pages won't work. Unfortunately, I have lessons again and Dr Wachsmuth wants a paper by 10 March, so there is a lot to do at once. If at all possible, I'd like to finish it this week. Perhaps you will then decide together with Dr Wachsmuth who will be the editor, whether both sections or neither. The difficulties are always the same. The natural sciences section is rightly responsible for agriculture, the medical section for medicine, and if both areas are covered in a paper, both must be represented. But it would be quite possible - after all, it is a small work - to have only the title: Physiological effects of the smallest entities and their consequences for medicine and agriculture, and then at the bottom, as in the case of the two writings Sternenwirken in Erdenstoffen: Experimentelle Studien aus dem Biologischen Institut am Goetheanum, L.K. Perhaps you can have a look at the two writings again.
> It makes no sense to bring the work in two parts; perhaps the medical part in Natura and the agricultural part in Gäa Sophia. The work is too small for that and loses its impact if it is torn apart. The "Clichés" for the silver book have turned out very nicely so far. Hopefully the book can be finished by Easter. With grateful admiration, L. Kolisko [253]

The letter to Dr Wachsmuth dated 5 February 1929, one day before the letter to Wegman, was as follows:

Dear Herr Dr Wachsmuth!

I apologise if I am a little late in replying to your request to write an article for the Agricultural Yearbook. I have been racking my brains incessantly as to what I could write for it. But you have set such a short deadline that I don't know how to get a solid piece of work ready for print. It's particularly complicated for me because another work, "Das Silber und der Mond" [Silver and the Moon], is currently at the printers and should also be published by Easter. Secondly, I have some work to do for the farmers' calendar, and thirdly, the work I promised in December on the Effect of the smallest entities and their consequences for medicine and agriculture is almost finished. I just have to make the final corrections and then the manuscript will be sent to you. And to top it all off, I have another main lesson in a senior class until 19 March.

However, you write that you intend to print my work from the last Gäa Sophia on the influence of the moon on plant growth. Perhaps it would be best if I extended this work by a year. Would you be happy with that? It could simply be printed, and in a second subsequent article, I would present beautiful results from the year 1927, i.e. an annual curve, as well as recent maize experiments which show that maize should only be planted during the full moon phase, but not during the new moon phase.

As far as potency work is concerned, you wrote to me in December that you would not be happy if it became a "Natura" number, as this would make it too medically orientated. I myself am not in favour of making a Natura number. It would certainly be better if it were a small, stand-alone brochure that doesn't cost a lot of money and is easy to hand out to anyone. Now you might think that the work for the yearbook could be used now. There would probably be an objection from the other side that it is too "scientifically" orientated. There is an urgent need for this little book to be published *quickly*.

I am being bombarded from all sides, including from members, that I should reply to Fenner and so on. There is also a definite desire to blur the priority of the work from the Biological Institute. For example, a few days ago I received a letter from a Nuremberg member, Dr Rissmann, in which the following is written:

"Since the dilution question has played an increasingly important role in science in recent times, including in our Erlangen Botanical Institute, there is a great danger in *Junker*'s presentation that anthroposophical research will be pushed to one side again and *Junker* will be presented as the father of potency dilutions. In passing, perhaps *Kolisko* is mentioned, who 'also' worked in this direction, but is not to be taken entirely seriously, since according to Junker (p. 662) she blindly ignores the details of her curves as the result of absurd cosmic speculation, etc. Moreover, the works are not easily accessible! [...]"

Now, of course, it is not possible for the work to fall on the floor between sections. I suggest either printing below it: published by the Medical and Natural Science Sections or simply publishing it without any reference as experimental studies from the Biological Institute at the Goetheanum at the Orient -Occident publishing house. You can then recommend the work to farmers and the medical section can recommend it to doctors. Perhaps this work could also be published at Easter. Please let me know if you agree with my suggestions. [254]

Ita Wegman replied to Lilly Kolisko almost immediately, on 28 February 1929. Despite Wachsmuth's increasing aversion to her and her work, she had no objections to the proposal of a joint publication, but wished for further publications from her for "Natura":

Dear Frau Kolisko, I have received your letter and the copy of the letter to Dr Wachsmuth. I have no objections to publishing your work together with the Natural Science Section. Dr Wachsmuth himself once said that this should actually be the case, but I don't know whether he still holds this opinion or whether he now thinks that each section should do its own thing. If that is the case, then there is no other option but for you to publish this as your special studies from the Biological Institute at the Goetheanum.

But now, of course, I would also like an article from you in "Natura", because you have been promising us this for so long. It would be so good if there were also an article in "Natura" on potencies in general. But of course, I also realise that you cannot write the same thing twice if you have already dealt with the potencies in your small paper, which is then to be published as an experimental study. Natura is very much a magazine that also circulates in welfare circles and is read by many people who have some kind of connection with medicine, without being too scientifically educated, so that one part can be given to the scientists and medical students, and the other to the medically interested. The problem with "Natura" is getting bigger and bigger because it seems that what I really intended as an idea cannot be realised. I always wanted to have this threefold structure: a medical journal for doctors, then "Natura" for people with a general interest in medicine to create a kind of medical movement, and then a series of publications in which more scientific work is done to satisfy the scientists. Despite all our efforts, we are not able to interest our co-workers in this kind of work. We don't get many contributions and we don't have the time or the opportunity to do everything on our own. And yet Dr Walter has fallen ill again, always with the same problems, worries about "Natura", as she says - of course there are other things involved - but she always says that this excitement about bringing the Natura articles together makes her so nervous and worried that it always makes her ill. And so Dr Walter will have to take another long holiday in the next few days to recover thoroughly. For me, of course, this means a loss of valuable help, but of course there's nothing else to do now. So I would be really grateful if time would allow you to send something to "Natura", as you often did in the past, just as I would like your husband to write something again, if possible. We have now used up everything we still have of his old work.

How are you all doing? Frau Werbeck is coming at the end of the week to give lessons here in the youth course. I very much hope to see you and Dr Kolisko before the General Assembly. With my best regards, dear Frau Kolisko.[255]

Silver and the Moon

The publication, "Silber und der Mond"[256][Silver and the Moon] appeared after Easter 1929 in the "Natura" series. The first part was an introduction with a historical overview of the effects of the moon and silver. In the second part, the connection between silver and the moon was demonstrated by means of experiments using capillary dynamolysis. Mention was also made of how the cycle of the year and the annual festivals are expressed in the pictures. Elisabeth Vreede wrote in a book review in the weekly magazine "Das Goetheanum":

> [...]Two rhythms interpenetrate: that the year, the sun, and that of the month, the moon. We can see the phases of the moon in the greater contouring and richness of form in the waxing as opposed to the waning moon, somewhat corresponding to the difference between the individual "day pictures" and the "night pictures. However, the effect of the sun during the year is revealed to us in a particularly characteristic way when we study it closely. What arises there in different seasons, changing from month to month, cannot be explained by purely physical causes such as an increase in temperature and the like, but true formative forces appear in the meaning that spiritual science gives to this word. The pictures from spring to winter are like germinating, sprouting, flowering, shooting into seed, then turning into fruit and withering away. What lushness of form in the May pictures, what torpor in November and December! The colours will also change. From the usual brown (with green) of the silver pictures to violet and colourful nuances in autumn, to light or dark green in winter. Through the rising and falling of the phases of the moon, sometimes intersected by solar and lunar eclipses, the character of the year can be clearly seen.

But then there seems to be a higher law, which goes beyond the purely physical and ethereal regularity of the formative forces, quietly announce themselves. This is where very special figures appear on the great feast days of the year. Easter, Whitsun, St John etc. bring something into the picture that does not belong to the usual rhythm of time. Something is expressed very delicately in the forms, sometimes also in the colours of the picture, as if something spiritual were quietly making itself felt in the material. Faced with the picture of St John with the stars (the symmetrical position in relation to the southern direction alone shows that no mere "coincidence" is at work here), one is led to the question that the author poses with ease: "Who imprinted the stars on the silver?"

Easter pictures are available from two consecutive years. On 12 April 1927, Easter Sunday coincided with the full moon; in 1928, Easter was three days after the full moon, which fell on Maundy Thursday. No one who really studies the pictures of the days in question will be able to escape the impression that a spiritual world is quietly beginning to speak here, as if in response to the faithfully persistent searching of the experimenter. The desire to learn more about the "workings of the stars in earthly substances", to learn to understand the spiritual language that has been expressed here by means of chemical solutions in the filter paper. From physical conditionality via etheric formative forces to spiritual-astral activity. [...] Whoever has been able to carry out such experiments systematically year in, year out, whoever has lovingly familiarised herself with what these pictures can say, has also acquired the right to point out the initially enigmatic phenomena that appear in individual pictures, especially on the distinctive days of the year.[257]

*

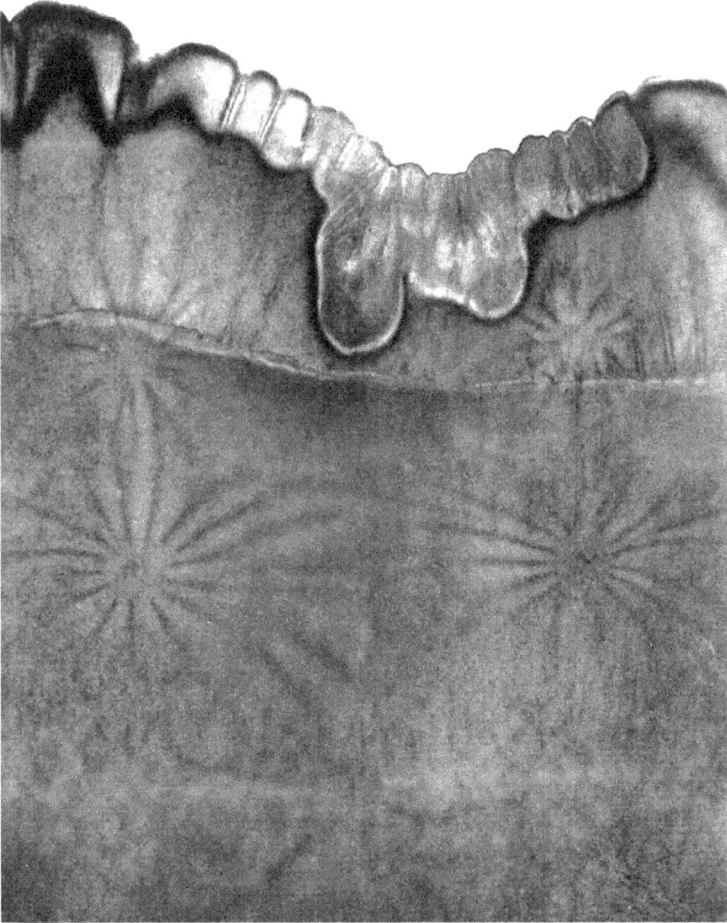

Fig. 34: St John's Day 1927. © Kolisko Archive, London.
From: "Spirit in Matter".

Meanwhile, Lilly Kolisko continued to correspond with Ita Wegman about the book discussed by Elisabeth Vreede and about the requested work on questions of potentiation. On 23 May 1929 she wrote:

> I am pleased that you are happy with the silver book. It looks quite nice, even from the outside. The Orient-Occident publishing house has really put a lot of effort into it. Now they must also sell a lot of copies. I have therefore decided to speak about silver and the moon right now in Hamburg. The conference is open to the public and perhaps even more people who have enough money to buy the book will be interested in it.
>
> You will certainly be surprised that the potency book is still not finished. But the fact is that I've had it ready since Christmas, but my husband found it too polemical and now wants to change every sentence. That takes a long time. He also wants to see a lot of experiments included in it, which are probably finished but which I still have to work on. For example, he would like me to publish a series of experiments up to the 600th potency with all seven metals and many other things. But that still means a lot of work. I had thought the scope of the paper would not be so broad. I am constantly working on it and ask for your patience. Miss Kreuzhage has been away from me since 1 May. A lot has to be changed at the Institute now until the work I have in mind is completed. If you haven't heard from me for so long, it's because there's so much work to be done.[258]

Also in 1929, Lilly Kolisko's "The Moon and Plant Growth" appeared in an issue of the "Gäa Sophia" magazine of the Natural Science Section, which was dedicated to agriculture. It included sowing experiments with wheat and maize during different phases of the moon. In summarising her results, she wrote: "[...] The only question that arises is: why do two phases of the moon produce two different results?

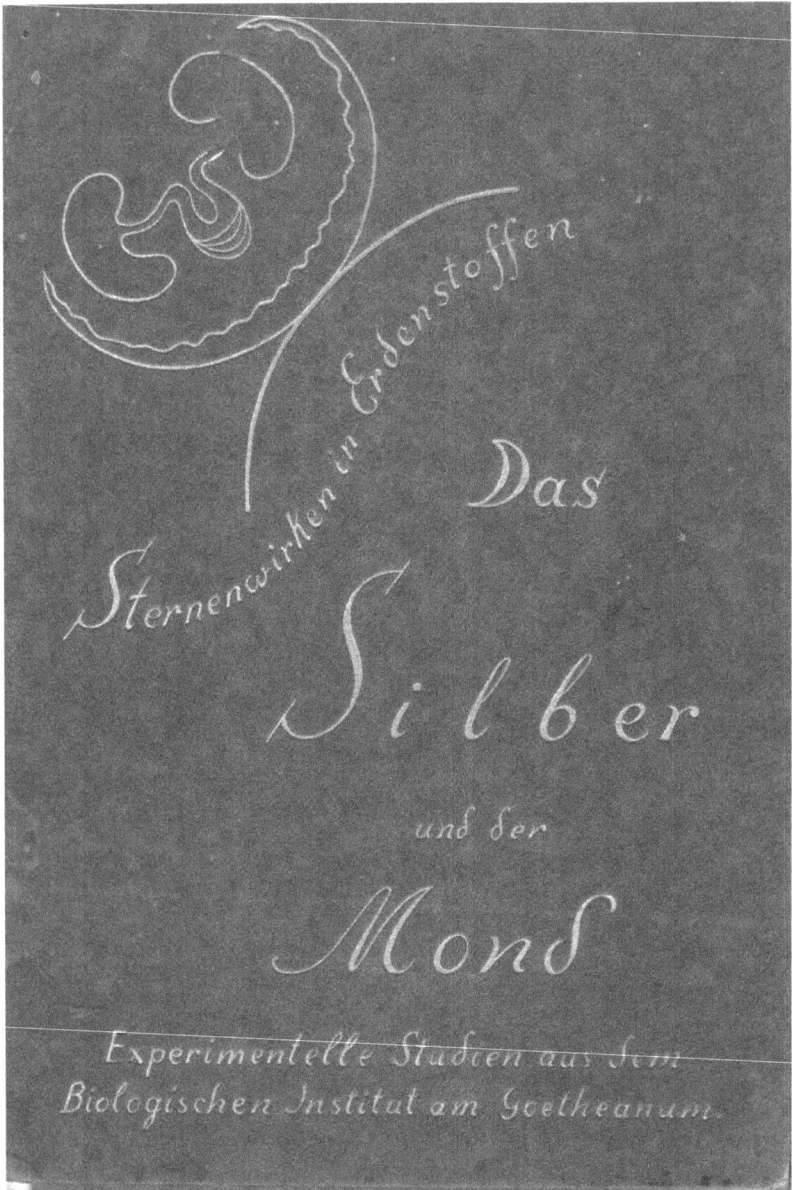

Fig. 35: "The Silver and the Moon", 1929. © Kolisko Archive, London

The answer is actually easy to find. If you want to supply a plant with the forces that flow in the earth at full moon time, then you must not sow the plant at full moon, but a little *before*. Then you bring it into the flow of the maximum rising moon phase. If you use the full moon day itself, then the plant will already receive *outflowing* forces, i.e. no longer those suitable for it [...]." [259]

Rudolf Hauschka

In November 1929, a significant conflict arose between Lilly Kolisko and Ita Wegman's new clinic co-worker Rudolf Hauschka - a disagreement that severely tested her trust in the co-operation with the Clinical Therapeutic Institute Arlesheim and the magazine "Natura", and indirectly also her relationship with Ita Wegman.

On 27. October 1919 Ita Wegman wrote:

> Dear Frau Kolisko! I heard through Stein that you are willing to write a short article on mercury [dilution] for "Natura". I received this news with the greatest pleasure and hope that you will have the opportunity to do so soon. Taking a short break from "Natura", we have now tried to put together a more popular magazine in which I will begin to present the importance of the earth and the bread that grows on it. However, I have to leave for England in the next few days and travel from there to Holland and would very much like to have the essays in print before I leave.[260]

From 8 to 10 November 1929 an important scientific conference took place in Amsterdam, Holland, organised by the Executive Council of the Dutch National Society. Among the speakers were the most important representatives of anthroposophical scientific work: Guenther Wachsmuth, Lilly Kolisko, Hermann Poppelbaum and Ehrenfried Pfeiffer. Rudolf Hauschka was also invited.

Fig. 36: Conference in Holland, November 1929.
Front row from right: Lilly Kolisko, Elisabeth Vreede, Guenther Wachsmuth,
Ita Wegman, Willem Zeylmans van Emmichoven. Behind Lilly Kolisko:
Rudolf Hauschka. Private property.
From: J. E. Zeylmans van Emmichoven: "Who was Ita Wegman",
Volume 2, Edition Georgenberg

After this conference, the issue of "Natura" mentioned in Ita Wegman's letter appeared with Lilly Kolisko's article "On Bread and Mercury" [261] - it dealt with the use of mercury as a seed dressing and the effect of small amounts of the substance on the human organisation. The results of her many years of research were clear: "[...] The Biological Institute at the Goetheanum [...] has carried out growth trials with mercury-containing seed dressings in various potencies, which will be published soon. It cannot therefore be said that the quantities are so small that they do not give rise to any health concerns, because the higher the dilution, the stronger the effect in the specific organ system [...]."

The same issue of "Natura" also published a longer reflection: "Bread and the Earth"[262] by Rudolf Hauschka, who at the beginning of 1929 had accepted an invitation from Ita Wegman - with whom he had been in contact for some time - to work in Arlesheim in order to set up new research laboratories for the production of remedies and dietary preparations at the Clinical Therapeutic Institute.[(263)] Hauschka, who had been involved with anthroposophy for many years and had spoken with Rudolf Steiner in 1924 about research into the etheric,[264] quickly familiarised himself with the now established procedures and developed his own methods. In contrast to Lilly Kolisko, who carried out experiments for years until she was convinced of the results achieved, Rudolf Hauschka began to publish his investigations in "Natura" as early as 1929.[265] To Lilly Kolisko's great surprise, he also published pictures in his article "Bread and the Earth" that were made with the help of capillary dynamolysis and were reminiscent of a previously unpublished work by her.

She felt deeply affected and immediately wrote to Ita Wegman (21 November 1929):

> Highly esteemed Frau Dr Wegman!
> Yesterday I received the latest issue of "Natura" and can't help but write to you about it straight away. The article by Dr Hauschka reports on experiments that are said to have been carried out according to the method of L. Kolisko. The moon-silver book is quoted. However, the moon-silver book does not publish any experiments with plant juices combined with metal salt solutions. The fact that I have been carrying out these experiments for years is only known to a few people who visit me in the laboratory and to whom I have shown such pictures

when it was unavoidable, as I do not like to discuss unpublished work. (I have had bad experiences with this.) When I show these experiments, I always add: "These are still unpublished works that go back to a task that Dr Steiner gave me, and I must ask that they not be discussed." Dr Hauschka has not yet been in my laboratory. Dr Suchantke was once sent to me by you so that I could show him my method of rising pictures. I clearly remember I also asked Dr Suchantke and Dr Tiefenbach, who was with him, not to talk about these experiments in particular, and not to carry them out themselves as I had to reserve the right to further elaborate and publish them.

The fact is, however, that Dr Hauschka has published such experiments in "Natura", which therefore originate from me and have not yet been published by me. In the outside world, there is a certain name for such an offence, which I probably don't need to tell you.

Apart from that, I am extremely sorry that experiments are being carried out in such an amateurish manner and are going out into the world under your direction. Perhaps you should bear in mind, if the tone I strike here seems a little too harsh, that I have worked out the method over many years of labour. Since 1924 I have been continuously engaged in exploring the forces that lie dormant in the plants in connection with the sun and moon, silver and gold and, of course, with all the other planets.

I have kept quiet about it until now because I didn't think it was time to talk about it. Now other people are going behind my back and publishing pictures that, from my rich experience, I have had to label as "failed" experiments that can simply be thrown away.

With the best will in the world, it is not possible for me to discover characteristic plant forms or even to see "animal" forms in Figures 1 and 2, nor can I find "lime forces stimulated" in them. Figures 3 and 4 also do not justify the conclusions drawn from them. The same applies to Figures 6, 7, 8, 9. In Figure 8, too, it is impossible for me to see anything "comparable to lung tissue".

Any serious scientist must reject such experiments and interpretations as amateurish. I'm sorry that I didn't know what would appear in this Nature issue earlier, because then I could have objected sooner. Under no circumstances would I have contributed to this issue myself. Now the disaster has already happened, and I will probably have to deal with it elsewhere. Working in this way also devalues the serious scientific work that is done in society.

Dr Hauschka wanted to see me at the laboratory in Stuttgart. Under these circumstances, I am unfortunately forced to ask him to refrain from doing so. You can only work if there is a certain basis of trust, and that has been completely taken away from me, both with Dr Hauschka and with the Medical Section.

I regret that it has come to this, but I would rather tell the truth openly than wage a hidden war. Yours sincerely, L. Kolisko [266]

17 years later, in 1946, Lilly Kolisko wrote about the capillary dynamolysis she had developed and practised:

[...] We are convinced that this method of Capillary Dynamolysis can solve many problems concerning the value of food. But it can only be handled by an expert who has been trained for years in the various fields connected with this subtle test. It is a very sensitive method that must not be abused in a fantastical manner [...].[267]

[...] A completely new and beautiful world opens for future research work. But we cannot emphasize enough: it demands a strict training and discipline. Since I have lectured for years about these subjects, I have watched how the public reacts when seeing such striking results. It happens often that the eye gets caught by some unimportant detail which appeals to the aesthetic sense. It has to be understood that *beauty does not always mean health.* I sincerely hope that I can make it clear on these few pages which are at my disposal: *For the interpretation of these experiments, we need sober judgment.*[268]

It is not known how Lilly Kolisko's conflict with Rudolf Hauschka and "Natura" ended.[269] A written reply from Ita Wegman is not found- it is very likely that she sought an oral dialogue, possibly in Hauschka's presence. In the following years, Lilly Kolisko and Rudolf Hauschka gave lectures at conferences of the Medical Section, and in July 1931 they spoke one after the other.[270] Lilly Kolisko's reservations towards him remained - and she never published in "Natura" again.

18.

Changes in the First Class of the School of Spiritual Science, 1928 - 1931

> [...] the mandate given to me by Dr Steiner still exists and is in no way affected by the decision of the Dornach Executive Council [...].[271]

E ven after Rudolf Steiner's death Lilly Kolisko held Class lessons at the School of Spiritual Science in Stuttgart with great care and seriousness. In 1928 she found herself in a difficult situation due to her close connection to various members of the Executive Council in Dornach. She always kept them informed about her work. The whole subject of the Class lessons was fraught with tension at the Goetheanum; the co-responsibility and thus co-leadership that Rudolf Steiner had given Ita Wegman for the First Class was neither understood nor respected in Dornach after his death.[272] The immediate cause of the controversy was a different one, however: when a Russian anthroposophist was arrested, the Class mantras were confiscated by the police. This caused great consternation and the question of who was responsible was intensively discussed at the Goetheanum and elsewhere, especially in Stuttgart. On 10 June 1928 Marie Steiner wrote to Lilly Kolisko:

> Esteemed Frau Kolisko,
> There is some excitement about the rumour circulating in Stuttgart that I was the one gave Mr van Leers the Class mantras for Russian members when he travelled there. I only became aware of the fact that this had happened later and could never have agreed to it if I had been asked for my opinion - Dr Wachsmuth and I knew that Dr Steiner had forbidden it. Every Class member had to return their Class card and mantras when they returned to Russia. Mr Steffen also only became aware of the matter later.

There is now excitement about this and, of course, because of my Russian connections, one could easily believe that I had been informed of the matter or had initiated it. Please convey the true situation to the members present in Stuttgart before the Class meeting by reading out this letter. I ask that no debate follow.

With the highest esteem, Marie Steiner [273]

Marie Steiner informed Albert Steffen about her letter to Lilly Kolisko, and Steffen agreed that Lilly Kolisko should read it to the Class members in Stuttgart.

Instead, Lilly Kolisko travelled to Dornach to discuss the matter there before informing the members of Marie Steiner's perspective through her letter. Marie Steiner did not receive her, unlike Ita Wegman and Albert Steffen. On 22 June 1928 Ita Wegman wrote:

Dear Frau Dr Kolisko,

After you left, I thought about everything for a long time, and a sudden and liberating thought occurred to me - that a solution might well be possible from your side. Now that you have spoken to the Executive Council in Stuttgart and, as we did together, you have once again explained everything of relevance to them in chronological order, perhaps it would be good to consider whether you might communicate these same messages to the members at a Class meeting - just as much as you consider necessary - and at the same time inform them of the letter you received from Frau Dr Steiner. Naturally, also present your explanations.

Alternatively - because apparently this chatter seems to be emanating from Stuttgart - might it be better to discover who has written these stories and then bring them together with the Executive Council in Germany and tell them what needs to be said.

These two things could be done. If this does not happen, then, as Steffen said, this letter will probably be printed in the members' newsletter, and then there will be nothing left but for me to give an explanation there as well, and then of course the Class will be completely exposed. Any Class meeting here in Dornach will always lead to a discussion, which should be avoided at all costs.

If the explanation comes from you, you could control everything firmly and not allow any discussion to arise.

Without taking sides, you could explain it all and remind the members of everything, all the admonitions that Dr Steiner gave in class, and also explain your appointment and, if necessary, everything which relates to me. However, the Executive Council must be fully informed, and you can make use of everything I have told you.

In the assembly you can of course be much more general, but of course you can also strongly emphasise that it is not only Frau Dr Steiner who is informed about the Class, but that I am just as fully in the picture and that I have sent help to my Russian friends taking every precaution, just as I have always seen Dr [Steiner] proceed with Russians who took part in the Class lessons in various cities where I was present. If you are strongly supported in this approach by the German Executive Council, a lot of good and fruitful things can come out of it.

It must also be emphasized that there is absolutely no reason to assume that the mantras have fallen into the wrong hands; one must also have faith that the spiritual world is still protectively watching over them, especially since one has experienced how effective they can still be for all those who participate in the Class with the right attitude.

I wanted to write this to you, dear Frau Kolisko, but of course I will leave everything to your free will as to what to do. You will feel inwardly what needs to be done. These were my thoughts after you left so that's what I have written. If you want to do something, don't wait too much longer, otherwise it can't be stopped.[274]

Lilly Kolisko then sent the following letter to Marie Steiner on 24 June 1928. to Marie Steiner:

Highly esteemed Frau Dr Steiner!
I am very sorry that it was not possible to have a conversation with you in Dornach on Friday, as I would have liked to discuss the matter mentioned in your letter with you in person. Unfortunately, I was unable to comply with your request to read the letter, as I am currently on a longer break from holding Class lessons. I have not called a separate meeting of Stuttgart

members for this purpose.

Naturally, I reject the rumours as untrue, which I have not yet heard, but I could not call a meeting and read out a statement that did not also bear the signature of Dr Wegman. Otherwise, it might easily appear that the members of the Executive Council who hold the Class lessons in Dornach are not in accord on this matter. This would greatly increase the difficulties, and I cannot be responsible for causing such anxiety among the Class members. I have told Mr Steffen this personally.

I very much regret not having been able to speak to you in Dornach and hope that this will be possible at another time. With the expression of reverence.

Yours faithfully, L. Kolisko [275]

It is not known how the matter progressed in detail.

*

However, the controversies in Dornach about the role of Ita Wegman and of the other Executive Council members in conveying the Class lessons, continued in full force.[276] On 19 February 1930, the Dornach Executive Council finally agreed to reserve all future deliverances of the lessons to Executive Council members - thus settling the internal disputes about the Class. The Executive Council informed the people who had previously read the lessons (and not just conveyed the mantra): Ludwig Polzer-Hoditz, Harry Collison and Lilly Kolisko. Lilly Kolisko received the following letter:

> Dornach, 10 February 1930
> Honourable Frau Dr Kolisko!
> To achieve consistency when holding Class lessons and to avoid further developments and difficulties, we ask that the holding of Class lessons, i.e. the reading of the texts of the Class lessons, be reserved exclusively for members of the Executive Council.
> As we have heard that you intend to resume Class lessons in Stuttgart, we would like to inform you of the Executive Council's decision. Yours sincerely[277]

Lilly Kolisko replied directly to Albert Steffen and wrote to him on 21 February 1930:

Honourable Mr Steffen!

I have been informed today by both Dr Wegman and Frau Dr Steiner that a resolution was passed by the Executive Council on 19 February concerning the holding of the Class lessons - which I unfortunately cannot recognize as justified. I have written a letter to both members of the Executive Council, a copy of which I enclose here for your information. In it, I request a meeting with the entire Dornach Executive Council, without which I consider a solution to be impossible. As this is a very urgent matter with serious consequences, I would also ask you to inform me of the date of this meeting as soon as possible.

Yours sincerely, L. Kolisko [278]

Her letter to Ita Wegman was as follows:

Honourable Dr Wegman!

I received your kind letter of 20 February and regret to inform you that, after careful consideration of the facts, I am do not consider the decision of the Dornach Executive Council as justified.

The right to read the Class lessons in Stuttgart verbatim from my own shorthand notes was granted to me by Dr. Steiner himself. I did not request Dr. Steiner to grant this but was commissioned by him.

My stance merely means that I will continue as I have since Dr Steiner's death. It is not a new matter concerning which I should have asked the Executive Council for permission. Whether the break between the last lesson and my resumption was long or short is not relevant. My decision to continue reading the Class lessons now has been strengthened by the many urgent requests I have received from Class members in Stuttgart.

I have already announced the Class lessons in Stuttgart and the surrounding area for 6 March, as there is a birthday celebration for all members of the General Anthroposophical Society which will take place on 27 February - the date I first announced the resumption. I cannot simply withdraw such an announcement.

The current situation, however, requires clarification.

This seems to me only possible verbally. I therefore request a meeting with the entire Dornach Executive Council as soon as possible and am happy to provide any requested clarifications then. Apart from Thursday, 27 February, I am available at any time and request a prompt response. A completely identical letter is being sent simultaneously to Frau Dr Steiner and a copy to Mr Steffen. [279]

The meeting with the Executive Council appears to have been organised at short notice - and Lilly Kolisko apparently asked for a few days to think about it afterwards. On 27. February 1930, Rudolf Steiner's birthday, she wrote to Albert Steffen:

Honourable Mr Steffen!

In view of the extraordinarily difficult situation for the First Class of the School of Spiritual Science in Dornach, I have decided to do the utmost in my power.

I will therefore not be reading any Class lessons for the time being. However, the assignment given to me by Dr Steiner still rightly exists and is in no way nullified by the decision of the Dornach Executive Council. I will immediately inform the members in wording about the Executive Council's decision.

Yours sincerely, L. Kolisko [280]

*

Despite Lilly Kolisko's very difficult decision the controversies within the Dornach Executive Council surrounding Ita Wegman and Class lessons continued. In response to an enquiry from Guenther Wachsmuth, which indirectly concerned her relationship with Ita Wegman, Lilly Kolisko replied on 23 March 1930:

Dear Mr Wachsmuth,

I apologise for only replying to your letter today, but I could not easily make up my mind to do so, as cannot understand why you cannot obtain clarification in Dornach itself. Nor does the point seem so important to me. It is not a question of whether or not I have informed Dr Wegman of my intention to read Class lessons again, but, as far as I have grasped the problem, it is a question of whether in future even more people should be authorised by the Executive Council to read the Class lessons or whether, on the contrary, everything existing should be dropped in favour of the Executive Council.

As you only want information from me for your own personal judgement, I will of course tell you what I said during the Executive Council meeting: During my presentation at the Agricultural Conference in January, I informed Dr Wegman verbally that I intended to continue with the reading of the Class lessons "soon", as many members were urgently requesting this and I could no longer ignore these requests. I did not specify a particular date at the time. Then I decided to begin again on 27 February, Dr. Steiner's birthday, and gave the same notice 14 days in advance as I always have.

I notified the German National Executive, which then had to inform the individual branches, the Free Anthroposophical Society and some members living in the Stuttgart area. As the break was somewhat longer this time, I also sent the message to Dr Wegman, which must have arrived in Dornach on about 15 February. I have not kept a copy of this letter but the content was approximately: "Dear Dr W[egman]! I would like to inform you that I decided, after a long break, to return to continue reading the Class lessons. The next Class Lesson will take place on Thursday 27 February".

This was my free decision, based on the situation in Stuttgart. I didn't discuss it with Dr Wegman beforehand, but neither did she tell me in January that difficulties might arise.

I found this form of notification sufficient, as I did not intend to ask for permission, but simply wanted to inform you that after a long break I am now continuing what I did on Dr Steiner's behalf during his lifetime. I hope that you can now see clearly and ask you to consider these explanations really are only for your personal information.

Yours sincerely, L. Kolisko" [281]

At the same time there were further conflicts in Stuttgart - Adolf Arenson, who was close to her and who apparently maintained his own study circle on the Class mantras and also wanted to read the entire Class lessons in future, [282] was described by Marie Steiner as the person in

Stuttgart whom Rudolf Steiner had appointed "for the esoteric work with members", corresponding to Harry Collison in England, Helga Geelmuyden in Norway and Anna Gunnarsson Wager in Sweden.

> "They are the people who were assigned this work by
> Dr Steiner himself." [283]

Marie Steiner distanced herself ever further from Lilly Kolisko - she and her husband had not followed her suggestions in 1925 and were far too close to Ita Wegman. Wegman, however, stood up for Lilly Kolisko and wrote to Albert Steffen on 7 June 1930:

> Esteemed Mr Steffen,
> In response to your letter, I feel compelled by duty to write you the following. Regarding Mr. Arenson's statements, I can, of course, only report those statements that were made to me by Dr. Steiner. These were, in essence, the following, very explicit statements: That only Ms. Kolisko could be considered for holding the Class lessons in Stuttgart.
> He gave her permission to take shorthand from the beginning and explicitly stated that he trusted her shorthand to be accurate. Since no other transcripts are available, because Ms. Finckh's shorthand only became known to him shortly before his death, I don't understand how this notification to Mr. Arenson to hold the Class Lesson should be interpreted. It's one person's word against another. I am still of the opinion, both before and after, that the Class must now remain solely in the hands of the Executive Council. Then we would have removed the Class from any discussion.
> Yours sincerely, Dr I. Wegman [284]

It is not clear exactly when Lilly Kolisko, with the approval of the Dornach Executive Council (or at least Ita Wegman) began holding the Lessons in Stuttgart again. On 22 January 1931, however, Ita Wegman wrote to Eugen Kolisko:

> Dear Dr Kolisko! I was planning to go to Stuttgart to talk to you and Mr Leinhas, and to your wife if possible. Unfortunately, I fell ill and had to stay in bed until today [...].

[...] After the Stuttgart meeting, Frau Kolisko should also hold Class lessons without further ado, if the members so wish, which will certainly be the case. Everything will go in two directions, nothing more can be done about it. The passage of time and positive work will ultimately create unity again. The duality will also exist in the Dornach Executive Council probably until there is intervention from above. If only work can continue and people can find each other in their work.

With very best regards, Dr I. Wegman [285]

19.

The last years in Stuttgart, 1931 - 1937, and the Burghalde Sanatorium

[...] The time is a very bleak, and no one knows what the future holds. I suspect that you, too, are carrying a heavy burden and have perhaps given up all hope of a change in the present situation [...].[286]

[...] And if we look at the events of our time that threaten to engulf us, then we can be filled with deep gratitude towards Rudolf Steiner, who was the first person to reveal to us the mysteries of the cosmos [...].[287]

[...] The work of Eugen Kolisko, Helene von Grunelius and Lilly Kolisko (in Unterlengenhardt) could thus be understood as an overture, in which everything that follows is already anticipated in a germinal way, but only gradually unfolds. This also applies to their collaboration, in which very different people with different impulses work together in such a way that they are active in the sense of a superordinate whole, learning from each other and complementing each other. [288]

The severe internal crises of the Anthroposophical Society, including the German National Society,[289] were compounded by the complicated political and economic situation in Germany which made all further work more difficult. Hitler's NSDAP was becoming increasingly powerful. In the summer of 1931, Lilly Kolisko was relaxing in Bad Aussee, Salzkammergut, Austria, in a holiday home that Eugen Kolisko had inherited.[290] In a letter that she wrote from there to Ita Wegman, she once again expressed her spiritual affinity, despite Hauschka and all the controversies surrounding the First Class:

Honourable, dear Frau Dr Wegman!

Yesterday would have been the occasion of my lecture on further research results on the smallest entities.

Due to the unfavourable political circumstances, the entire event had to be cancelled. I would like to say that I regret this in every respect.

The time is a very bleak, and no one knows what the future holds. I suspect that you, too, are carrying a heavy burden and have perhaps given up all hope of a change in the present situation

You will probably find it strange that I am writing to you after we have had nothing to say to each other for so long, but I would like to make one thing very clear: No matter how great the differences may be in the factual assessment of the situation and the resulting actions, I will always feel inwardly and spiritually connected to you and be prepared to support you to the best of my ability in difficult situations.

As far as the lecture prepared for the Medical Section is concerned, I am of course prepared to present it at any time if you can hold the conference. I was already thinking about Michaelmas, as the emergency order ends on 1 October, but that falls during school term time and my husband would not be able to come.

Yours sincerely, L. Kolisko[291]

The conference at which Lilly Kolisko wanted to speak was the medical conference at the Goetheanum announced for 5 to 30 July 1931, which had to be cancelled at short notice. Ita Wegman replied more or less immediately on 7 August 1931:

My dear Frau Kolisko!

I always intend to write to you, I even had a letter half finished, but things have changed again in the meantime. Your kind letter made me extremely happy. We were also very sad that the medical conference could not be held. When the serious news came from Germany, I couldn't help but think that chaos would soon break out. Actually, the chaos is probably already here - it's just that it hasn't worked out as we thought it would. In some way there still seems to be a small window of opportunity in which something can perhaps still be realised. The fact that along the failure of the Anthroposophical Society also the world situation has become extraordinarily difficult is clearly noticeable. The only question now is: what else can we do? - Probably not to make things better, but to save something for the future. I only see the possibility of spiritual work more and more in the 'islands of culture' of which Dr Steiner spoke, because I no longer believe that anything can be done from the Goetheanum to prevent a collapse or to solve the questions connected with the collapse. It's all just theory and more theory, big words and small deeds. And when you then consider that Boos is trying to solve this depressing question related to the world situation, you can't suppress a tremendously bitter feeling. We can no longer do it from here, and we must try to find a centre somewhere else. We already want to hold our social conference, which we want to hold for the welfare workers, not here, as we intend to do, but in Berlin. It doesn't look as if the border will be free here, and there's no point holding this conference for anthroposophist's, other people have to be interested in it.

We could perhaps hold the medical conference here in Dornach in October, at least the discussions are going towards possibly doing this, although I will be saddened again that your husband would not be able to take part.

How about organising a meeting one day? For example, Gnadenwald, during the month of August, to talk about the situation.

Hope you and Geni are both doing well and recovering well. [292]

However, Lilly Kolisko had to return to Stuttgart and then to Dornach and had neither the time nor the money to travel to Gnadenwald. So, she replied to Ita Wegman from Bad Aussee on 14 August 1931:

Esteemed Frau Dr Wegman!
I have received your kind letter, but unfortunately, I don't see any possibility of travelling to Gnadenwald. It would mean a fast train journey of about 9 hours. Moreover, I must give a lecture for Dr Vreede in Dornach on 6 August. I will therefore have to travel from here to Stuttgart on 21 August to make the necessary preparations and then come to Dornach on 26 August. My husband and Geni will stay here until about 7 September. If it is convenient for you, I can visit you while I am in Dornach to discuss what is necessary.
Yours sincerely, L. Kolisko [293]

Collaboration with Elisabeth Vreede
in the Astronomical Section

At the Astronomical Conference at the Goetheanum at the end of August, Lilly Kolisko spoke about "Jupiter and Tin". Elisabeth Vreede wrote in her conference report:

As we know, Frau Kolisko has long been investigating the <u>mineral world</u>, especially the so-called planetary metals, in their connection with the cosmos. In this work she has proceeded quite systematically. A publication on "The Silver and the Moon" has been available for some time. It was therefore extremely important to be able to follow the next step, which illustrates the workings of Jupiter in tin. Numerous slides allowed one to experience for oneself how well-established is the connection, for example, of Jupiter with tin, of the Moon with silver. [294]

At the same conference, Joachim Schultz spoke about his experiments at the observatory at the Goetheanum on the influence of the stars on plant growth. He presented his method of exposing germinating seeds to the twelve different effects of the zodiac.[295]
For the "Kalender, Ostern 1931 - Ostern 1932" published by Elisabeth Vreede, Lilly Kolisko wrote the article "Gedanken zu bevorstehenden

Kalenderreformen" [Thoughts on forthcoming calendar reforms]; as was already the case during Rudolf Steiner's lifetime, endeavours were underway to fix the date of Easter (the first Sunday after the spring full moon), for practical / economic reasons. Through her research, Lilly Kolisko had gained deep insights into the cosmic foundations of the annual festivals and wrote in the introduction to this essay:

> In last year's calendar, Dr E. Vreede's article already referred to moves to set Easter on a specific date. We are thus facing a significant event, and if one believes one recognises the significance of this to any extent, then it may well seem justified to comment on it in more detail.
> Easter is part of the course of the year in a very special way. It is truly a cosmic festival, which establishes itself through the sun and moon. It stands before us as a witness to mankind's former deep connection with the cosmos. The feast of the resurrection of Christ, the Spirit of the Sun, who himself became man in order to redeem mankind, could not be arbitrarily determined. It always had to point to the great cosmic deed through its dependence on the position of the sun and moon. The feast of the resurrection was celebrated on the Sunday after the first *full* moon of spring, based on ancient starry wisdom. The sun and moon first had to enter into a certain relationship with each other before this festival could be celebrated. The sun and moon do not mean as much to modern people as they once did to those who were able to read the heavenly script. The souls of men used to be gripped in their deepest depths by the Easter mystery when they looked up at the full radiant disc of the moon after the sun had passed the equator on 21 March. People experienced a world mystery at the full moon of Easter. They can no longer do that today [...].[296]

She went on to write about the changes to the annual festivals proposed in the "Karlsruher Tagblatt" of 10 January 1932, among others:

> [...] This proposal is a truly excellent piece of work. Not only Easter, but also Christmas is to be cancelled. The winter solstice, Christmas and New Year are to be

merged into one big three-day festival.

With a light hand, the deeply meaningful mystery of the thirteen holy nights is also wiped away. Such suggestions can be made today because people have really forgotten the true spiritual meaning of the festivals. They can no longer experience the ancient wisdom, and there is no longer any knowledge of it today.

And if we look at the events of our time that threaten to engulf us, then we can be filled with deep gratitude towards Rudolf Steiner, who was the first person to reveal to us the mysteries of the cosmos. Without him, we would not know what a tremendous event takes place on New Year's Eve, for example. The entire plant world experiences something different on New Year's Eve than on the other nights of the year. Two streams of consciousness that run separately throughout the year intersect on this night. The mineral and plant consciousness interpenetrate and the plant world receives the power to blossom and bear fruit in the coming year. The point in time at which this powerful event is fully realised cannot be determined by humans. Man may arbitrarily determine Easter and New Year, but the starry world is not determined by man. Something happens spiritually in the 13 Holy Nights. However, it cannot be grasped with hands, determined in the usual sense on the scales with the weight or measured with the yardstick. Nor does it bring any scientific advantages - it can only fill the human soul that lives through this time with understanding, with devotion and holy awe.

And yet - even though it is not so easy to grasp what goes on in the world during the *true* festive seasons - an intimate study of nature can eavesdrop on the secrets of the world. If we penetrate deep enough into nature, it leads us to the spiritual.

At the "Biological Institute at the Goetheanum" (in Stuttgart), experiments have been carried out for years to find out the connection between the world of stars and the world of matter. Beautiful experimental results have already been published (Sternenwirken in Erdenstoffen I, II, III). In the most recently published book through many pictures, "Silver and the Moon" describes a substance that has always been associated with the moon. This is shown through the forms of a water-soluble salt (silver nitrate). The silver solution is allowed to rise up filter paper and be exposed to sunlight. The sunlight "reduces" the silver - in scientific terms - or, to put it another way, the silver, the earthly representative of the moon, enters into a relationship with the sun over the course of the day. At night, the process continues without the influence of external light. A picture is created on the paper, a different one every day. The relationship between the sun and moon is also different every day. If you then study the pictures for a whole year, you can read the course of the sun and moon and also the *holy days* from the pictures. These explanations are based on extensive experiments which have not yet published. It only points to the possibility of grasping the reality of the festive seasons even in the natural world. The scientific explanation will be given elsewhere.

The most wonderful experience in this respect can be had when looking at the silver pictures at Easter, Pentecost, St John's Day, Michaelmas and Christmas. These festivities are points of reference over the year. We include here the silver image on Easter Sunday 1927, Holy Saturday 1928, Easter Sunday 1928 and an "ordinary" picture from the course of the year. A scientific experiment has been carried out, and it is up to the viewer to read as much as he can into the results of the experiments. Everyone will probably find something that speaks to their soul and spirit about the powers of resurrection.

The whole universe is permeated by the forces of resurrection. Not only the plant world feels the awakening call of the sun and moon, the mineral world also hears it. [297]

Crystallisation experiments

During the winter of 1931/32, Lilly Kolisko continued the crystallisation experiments she had been working on for the previous ten years. In 1934 she explained in a publication:

Once again, the research began with a suggestion from Dr. Rudolf Steiner. The idea was to study the influence of plant saps on the crystallization process, how the plant-forming force impacts the crystal-forming force. Anyone who wants to pursue this suggestion must first acquire knowledge of these two cosmic forces, of the crystal-forming force on the one hand, and of the plant-forming force on the other. I sought to acquire this knowledge in various ways. If one does not have this knowledge, one can never distinguish when one force overcomes the other. [298]

These laboratory experiments were carried out by Lilly Kolisko: five different salt solutions were poured into dishes every hour for 24 hours, and the crystallised salts were weighed after twelve hours. The entire process took a total of 36 hours until the last sample was ready. The experiments showed that crystallisation took place differently during the night than during the day. The influence of different concentrations and different temperatures of the salt solutions was also investigated.

In the following years from 1931 to 1934, she also carried out her underground crystallisation and plant experiments on a daily basis, at depths of up to 16 metres under the surface. Her summary of this work confirmed Rudolf Steiner's statements on the crystallisation forces over the year: "[...] We must try to understand these two forces in nature: the mineral-forming force and the plant-forming force. Our experiments with wheat plants carried out for the purpose of studying the influence of the moon, show that maximum growth appears between June, July and August. That is just the period when the forces of crystallisation are at their minimum. When the plants are growing and growing, matter cannot contract and form heavy crystals. [...]."[299]

She also studied planetary influences on crystallisation. In "The agriculture of the Tomorrow", she gave important advice on how to carry out the trials:

"It is not advisable for the study of planetary influences (conjunctions or oppositions) to prepare the solution in advance and then only pour it into the dishes for crystallising. It is important to dissolve the salt at the exact times and then put it into the glass dish for the crystallisation. A careful observation will show that the salt also *takes a different length of time for the dissolving process* during a conjunction or opposition."[300]

Jupiter and tin

Working on a compilation of her various publications took a lot of time and energy. And unfortunately, her loyal colleague Wilhelm Kaiser was ill for a long time. He was the one who had always produced the pictures for her publications.

At Easter 1932, Lilly Kolisko then wrote the foreword for the fourth volume in the series of the Mitteilungen des Biologischen Instituts am Goetheanum: "Der Jupiter und das Zinn"[301]. This time the volume was published by the Mathematical-Astronomical Section at the Goetheanum. It contained her experiments with tin chloride and other metal salts during various Jupiter constellations in the period from 1928 to 1931. The publication was immediately translated into English.[302] In the procedural details, Lilly Kolisko gave an advice that became important for all those who wanted to reproduce her experiments. Regarding the use of the tin solution for capillary dynamolysis, she wrote, among other things:

"[...] We used stannous chloride ($SnCl_2$= tin chloride) for all our experiments. It is produced by dissolving tin in hydrochloric acid. Tin chloride is easily soluble in water. The aqueous solution has a strong reducing effect; it absorbs oxygen from the air and partially forms basic chlorine if the liquid is not too acidic. For the experiments, it is important not to use a fresh tin chloride solution, but to always have a 14-day-old solution ready that has already absorbed oxygen from the air."[303]

England

In England, biodynamic work was attracting growing interest. Ita Wegman wrote to Eugen Kolisko on 14 December 1928, after a trip to London and Brussels:

> [...]Mrs [Marna] Pease, who now has a beautiful flat in London, has a very great influence in this Executive [of the English National Society], not only by her beautiful harmonious personality, but also by the economic and social position she occupies. She has now furnished her house in London in such a way that people are always able to live there who want to do something for Society in England. - They very much want to promote agriculture because various farm owners in England have already joined forces and Mrs Pease is also very enthusiastic about this section and has also become very involved with it. Together with Mrs Merry, she has bought a small country house near London with quite a lot of land where they can start a kind of market garden. They also want to give Dr Mirbt a place for experimentation where the other farmers can meet and get advice. If this is managed properly, perhaps some things can be organised from there [...].[304]

Eugen Kolisko had already started monthly medical consultations in London in 1925 at Ita Wegman's request, as had Willem Zeylmans van Emmichoven. In the autumn of 1925 the book by Steiner / Wegman "Grundlegendes für eine Erweiterung der Heilkunst nach geisteswissenschaftlichen Erkenntnissen" [Spiritual-Scientific Fundamentals of Therapy] had been published in English in London. Ita Wegman had built up contacts with doctors there since the beginning of the 1920s and had seen patients and spoken to colleagues with Rudolf Steiner in London in 1923 /24.[305] George Adams had translated the Agricultural Course, and Lilly Kolisko spoke at various "summer schools". Rudolf Steiner had often mentioned her work in his medical lectures; so, she came to England again in 1932, this time for the "Annual Meeting of the Agricultural Association", organised by the Anthroposophical Society in England and the "Anthroposophical

Agricultural Foundation".[306] She gave lectures between 7 November and 4 December 1932 at Rudolf Steiner House in London and in Birmingham. The first two on 27 November in London and the one on 28 November in Birmingham were entitled: "Influence of the Stars on Earthly Substances: Jupiter and Tin" and followed on from her latest publication. During the Annual General Meeting of the Anthroposophical Agricultural Foundation, she reported on her latest research at the Biological Institute at the Goetheanum ("The Work of the Biological Institute"), on the influence of the "smallest entities" on plant growth and animals (experiments with white mice). In another lecture she spoke about the "Influence of the Moon on Plant-Growth and its Significance for Agriculture". There was also an immersive exhibition of original pictures of the experiments with capillary dynamolysis, and the opportunity to ask questions.

Lilly Kolisko gave a further lecture for the anthroposophical "Human Freedom Group" at Rudolf Steiner House on 30 November 1932 about her discoveries on the "Cosmic Influences on the Four Annual Festivals". In it, she lectured on Rudolf Steiner's imaginations of the seasons. Carl Alexander Mirbt - who set up biodynamic agriculture in England together with his wife Gertrud (at the request of Daniel Dunlop) and worked in conjunction with the curative education institute in Clent - presented Lilly Kolisko's findings:

> [...] Using the same metals and the same process, completely different pictures are obtained during Easter, Christmas, St John's Day and Michaelmas. In connection with each annual festival, Frau Kolisko quoted a poem by Rudolf Steiner and was able to show the truth and reality of these imaginations in her pictures. Two of the pictures, Easter and Christmas, were reproduced as special colour prints. It is a wonderful experience to compare the picture made during Easter in the daytime with the picture of Christmas at midnight. In the first, rejoicing over the resurrection of Christ, in the other, the child, protected by the Virgin's mantle.[307]

His report on events with Lilly Kolisko continues:

> We look back on this festive week with gratitude, but

our feelings are mixed with slightly envious questions: How can we always attract such a large audience for our meetings? Many new faces were there, and among them many young people. Is a new visit by Frau Kolisko - which we are eagerly looking forward to - the only option? Or could we try to establish experimental scientific work here in the country as well?

Her London lectures were also noticed by the public; "The Daily Mail" (of 2 December 1932) and "The Observer" (of 4 December 1932) published predominantly positive assessments.

Further publications: "Physiological proof of the efficacy of the Smallest Entities" and "The Moon and plant growth"

At the beginning of 1932, Lilly Kolisko was able to complete another part of her potency work, which had long been requested by the doctors. The publication of "Physiological Effect of the Smallest Entities" was organised by the Medical Section at the Goetheanum[308].

In the spring, a research report of particular importance to farmers and gardeners appeared in the "Mitteilungen des Biologischen Institutes am Goetheanum": "Der Mond und das Pflanzenwachstum".[309] In addition to the sowing experiments with wheat, Lilly Kolisko reported there on her extensive experiments with maize and various vegetables - sown during different phases of the moon - which she had carried out repeatedly in the years 1925 to 1932. In the end, the trials were carried out outdoors, and the yield, flavour and health of the plants were important criteria for the assessment. For example, she wrote about kohlrabi:

> "In terms of quality, the full moon kohlrabi were juicier, more tender and sweet-tasting - the new moon kohlrabi were drier and slightly woody." [310]

The yields of the vegetables sown two days before the full moon were significantly higher than those sown two days before the new moon.

Lilly Kolisko also reported that the work she had published in "Gäa Sophia" in 1929 had inspired some farmers to carry out their own trials. She cited two specific cases of trials with maize in Piura in South America and in Kenya (East Africa). There, a farmer confirmed that he had achieved a 30 to 40 % larger yield. during the waxing moon (compared to the waning moon). The biggest differences were visible one to two days before the full moon, compared to one to three days before the new moon.

Situation in Germany from 1933

The political situation became increasingly threatening. In a letter from Walter Johannes Stein to Ita Wegman dated 8 May 1933, one can read that the biodynamic work and that of the First Class of the School of Spiritual Science were both at risk:

> "[...] Last week the National Socialists were in Berlin at the Anthroposophical Society and asked whether it was a secret society. They demanded the Agriculture Course. They were referred to Mr Bartsch. I warned Kolisko and Rittelmeyer. In any case, Frau Kolisko was advised not to hold a Class Lesson [...]." [311]

The Waldorf School also received visits from the authorities. Lilly Kolisko wrote regarding her husband's situation:

> "At the request of his colleagues, Dr Kolisko had written a suggestion for the first lessons in chemistry. These suggestions were first published in 1932 in the bi-monthly journal 'Zur Pädagogik Rudolf Steiners'. They were later published as a special edition by the publishing house 'Freie Waldorfschule' and displayed in the following form: The aim is to introduce the basic concepts of chemical processes to the child. These must be concepts that can grow with the child. The child must then be able to take them with them through the entire course of the chemistry lessons, constantly expand them and always be able to confirm them anew with the experiences they gain. This book was attacked from one

side or another, and a commission of enquiry came to the school to examine the lessons. This commission gave a negative judgement. That was to be expected; how could a National Socialist commission of enquiry approve of teaching according to Rudolf Steiner's pedagogy?" [312]

"Teaching now had to be different. This was particularly so in the teaching of *anthropology* [Menschenkunde]. Certain things had to be taught to the children about the theory of *heredity* and about *racial studies*, especially in the upper classes and in the classes held in preparation for the Abitur. Dr Kolisko could not compromise. He could not teach things that ran counter to Waldorf education." [313]

However, Eugen Kolisko and his work were not only criticised by the Commission and the state authorities. [314] but also by many members of the teaching staff, who were drawn into the social controversy surrounding Ita Wegman and her friends, to whom he belonged. [315] "In addition, there was the *schism* that ran through the Waldorf School staff, as well as through the whole Anthroposophical Society. This was the case in every institution. The human relationships had changed fundamentally over the last few years. The basis of trust had been destroyed. Whenever you spoke to someone, you had to ask yourself which side or party they belonged to. You had to be very careful not to risk receiving another letter of accusation or a phone call to Dornach. Nobody can stand this situation for long." [316] To the horror of the schoolchildren and countless parents, Eugen Kolisko finally decided to leave the Waldorf school in the spring of 1934. [317]

Economic challenges for the continuation of research work

The financial situation of the Biological Institute at the Goetheanum had also been precarious for years, even though Ita Wegman donated a monthly contribution from November 1931 - as did several other people:

Dear Frau Kolisko, I have asked Mr Leinhas to send you Fr. 50 per month for your work in the laboratory.

This contribution should have been made available to you in November and December and now in January: I don't know whether Mr Leinhas has done so. This money would have been sent to me and can, instead be sent to you for your work ..."[318]

Lilly Kolisko had to continually seek new ways to raise the minimum amount of money needed to maintain her research and publish further publications. In 1932, out of necessity, she had begun to print her pictures of capillary dynamolysis for sale. They were apparently also distributed via Ita Wegman's clinic; on 16 December 1932, a letter from Lilly Kolisko to Wegman's colleague Madeleine van Deventer, who had been in the circle of "young physicians" with her, stated:

> Dear Frau Dr Deventer!
>
> I hereby acknowledge receipt of your check, which made me very happy. I hadn't even considered that they would bring in any money. I also think it's very nice that you've sold half of the uranium pictures. Perhaps the other half will also be sold by Christmas. I'm enclosing some new pictures for you. Gold for Easter and Gold for Christmas. Perhaps you'll find a buyer for those as well. Please take one copy each for yourself and for Dr. Wegman. We are selling them here for 0.60 M. each.
>
> Book stands receive a 20% discount. Would you be so kind as to give the enclosed photograph to Dr. Schmiedel? He needs it for the Weleda in-house magazine. But it's urgent. With many thanks in advance and best regards, L. Kolisko [319]

However, the sale of pictures could not, of course, support the Institute. In autumn 1934, when the exclusion of Ita Wegman from the Executive Council and the leadership of the Medical Section as well as the exclusion of countless members from the Anthroposophical Society was in preparation - six months before the decisive General Assembly - the Goetheanum ceased all further support for the "Biological Institute at the Goetheanum", which Rudolf Steiner held in such high esteem. At the beginning of September 1934, Lilly Kolisko received the following letter:

Honourable Frau Dr. Kolisko,

We regret to inform you that, as of 1 October, the Goetheanum will unfortunately no longer be able to continue sending the monthly amounts of 170 Marks to the Biological Institute due to the general economic situation. As with other businesses, we are forced to impose severe restrictions on ourselves. We would like to inform you of this so that you can prepare accordingly.

With highest regards, P. General Anthroposophical Society, Goetheanum Building Administration Department, signed Aisenpreis. [320]

Lilly Kolisko wrote in her sober reply:

Honourable Herr Aisenpreis!

I regret to learn from your letter of 31 August that from 1 October this year you will no longer be able to give me any support for the Biological Institute at the Goetheanum in Stuttgart apart from the amount I receive each month on the basis of the contract for the sale of the K. T. publishing house to the Philosophical Anthroposophical Publishing House. I must now see how I can continue the work with the help of voluntary contributions from friends who have already supported me considerably in recent years. These friends have now formed an association to support the Institute. With the highest consideration.

With highest regards for the Biological Institute at the Goetheanum, L. Kolisko. [321]

Friends of Lilly Kolisko's work in Germany did indeed try to help. In October 1934, for example, Paul Gimmi wrote a letter to the members of the Anthroposophical Working Group - the organisation that had formed in various countries following the progressive marginalisation of Ita Wegman and Elisabeth Vreede by the dominant Dornach Executive Council faction (Albert Steffen, Guenther Wachsmuth and Marie Steiner), in Germany mainly around Eugen Kolisko in Stuttgart.

As you can see from the enclosed copy of a letter from the Goetheanum, the situation of the Biological Institute is very difficult. Due to the "general economic situation", the Goetheanum is unable to continue to send the support of Mk. 170 per month that it had previously granted to the Institute. This contribution has already been cancelled this month. You will also see from the letter that the sum of 200 marks per month from the "publishing contract" is all that is available.

This contract, which was concluded at the time between the K.T. [Kommenden Tag] and the Philosophisch-Anthroposophischer Verlag, expires in mid-1935. This means that from 1935 the Biological Institute at the Goetheanum is without financial means and can only continue to exist thanks to friends who are interested in the continuation of the work. As the current economic situation is generally very difficult, we saw no other way of ensuring the continued existence of the Biological Institute at the Goetheanum in Stuttgart than to found an "Association for the Promotion of the Biological Institute at the Goetheanum in Stuttgart".

If you are interested in ensuring that this work can continue, we cordially invite you to join this association as an associate member. We have set the monthly dues at least 1 Mark and, in return, we will provide the "Mitteilungen des Biologisches Institut am Goetheanum" [Announcements of the Biological Institute at the Goetheanum] free of charge. Issue 1 was already published in June; we expect to be able to print Issue 2 before Christmas. However, we hope that some friends will be able to subscribe to more than the minimum monthly contribution of 1 Mark. We ask that you let us know immediately whether we may send you a membership card. Sincerely yours, Gimmi.[322]

The "Verein zur Förderung des Biologischen Institutes am Goetheanum in Stuttgart" was founded on 6 October 1934. The founding letter was signed by Emil Leinhas, Clarita Berger, Blickle, Hoh. Berner, W. Fink, Paul Gimmi and Hans Kleemann.

Announcements of the Biological Institute at the Goetheanum

During this time, Lilly Kolisko compiled the series of publications entitled "Mitteilungen des Biologischen Instituts am Goetheanum".[323] Booklet no. 1 had already appeared in June 1934, no. 2 was published at Christmas 1934; no. 3 and no. 4 appeared in the course of 1935. In these booklets, Lilly Kolisko published some older but as yet unpublished research results as well as her latest findings. In the medical field of "Smallest Entities", for example, she wrote about her comparisons of lactose and alcohol as potentising agents; from her research on the biodynamic preparations, she used capillary dynamolysis to show differences in quality between the urine of cows, horses and deer. For gardeners and farmers, she also set further accents with her reports on sowing experiments in the garden during different phases of the moon. In the astronomical field, her investigations with metal salts during various constellations and during the meteor shower on 9 October 1933 were enlightening. Under the title "Gestalt forces in the plant world", she reported for the first time on the relationship between plants and metals using capillary dynamolysis, the application of which for the diagnostic examination of human urine was discussed in detail in the fourth issue. She had already published some of the results of her many years of studies on crystal-forming powers in issues 1 and 2. She then published a small folder of postcards with 12 crystallisation pictures under the title "Kristall-Gestaltungskräfte I" [324]

37: Lilly Kolisko in the laboratory. From: "Modern Mystic", February 1939

Sanatorium Burghalde

After leaving the Stuttgart Waldorf School, Eugen Kolisko looked for a new location for his medical practice. With the help of Clarita Berger and Ottilie Matthiessen, the "Sanatorium Burghalde" was opened in Unterlengenhardt on 1 April 1935. Eugen Kolisko and Helene von Grunelius worked there,[325] and so the Burghalde quickly developed into a flourishing place, despite critical observation by the National Socialist authorities.[326] Lilly Kolisko also worked partly in Unterlengenhardt:

> Lilly Kolisko was probably not formally employed at the Burghalde. Nevertheless, she continued her research there, including investigations using the rising picture method. A special form of co-operation between Lilly and Eugen Kolisko and the practical value of Lilly Kolisko's research has been handed down to us by Margot Rössler. She noticed that the patients received their injections at different times of the day. One night after midnight, Eugen Kolisko appeared in her room with a doctor's coat and injection syringe in his hand: "My wife has calculated that this remedy has the best effect at that moment." Here it becomes clear how exact experimental results obtained through immeasurable diligence, accuracy and consistency had a therapeutic impact. A task not yet fully recognised even today...[327]

In the early hours of the morning, Eugen Kolisko was seen in the meadows picking flowers for his wife, which she needed for her investigations in the laboratory.[328] During the time at Burghalde, Lilly Kolisko was intensively involved with the production and effects of the bio-dynamic preparations, including their shelf life.

Eugen Kolisko also used the jointly developed diagnostic methods at Burghalde, for example in the medical analysis of human faeces. Lilly Kolisko wrote in "The agriculture of Tomorrow":

> [...] A long and careful study is necessary to be able to judge with absolute certainty, whether the excretion comes from a perfectly healthy person, or whether

pathological changes are present. Each individual has a characteristic formation. But it is possible to use our method of Capillary Dynamolysis to diagnose in the excretions various diseases. We carried out these observations for more than fifteen years, with many healthy and sick people. Dr. Kolisko, as school doctor in the Waldorf School, Stuttgart, cared for about 1,200 children and made liberal use of this method for diagnosis; as he did also later on, in his clinic in Burghalde. From this experience we can state that it is not only possible to give the correct diagnosis for various diseases, but we can also follow up the healing process. Capillary Dynamolysis is a very subtle method, which can show the beginning of a disturbance in the equilibrium of forces very early, much earlier than is possible with a chemical or microscopical test. The equilibrium of the forces is disturbed before we can see it in actual disease; that means it is possible to prevent the outbreak of a disease in time, and to continue the treatment long enough to ensure a real cure by watching the gradual recovery of the lost equilibrium.[329]

Developments in the Anthroposophical Society and the School of Spiritual Science in 1935

By resolution of the General Assembly of the Anthroposophical Society on 14 April 1935, after years of agitation against them, Ita Wegman and Elisabeth Vreede were ousted from their positions on the Executive Council and in the Sections.[330] Leading individuals and working groups associated with Ita Wegman and Elisabeth Vreede who had largely forged autonomy as a result of the events of the previous years (the "United Free Anthroposophical Groups") were excluded from the Society including Eugen Kolisko and Jürgen v. Grone in Germany, Daniel N. Dunlop and George Kaufmann in England, Frederik Willem Zeylmans van Emmichoven and Pieter de Haan in Holland.[331] Unlike her husband, Lilly Kolisko's membership in the Society was not revoked. However, various allegations were made about the beginnings of her involvement in Class Lesson in connection with accusations against Count Polzer-Hoditz. He had spoken against the dismissal of Ita Wegman and the whole development at the General Assembly[332] and was now called to account by the Dornach leadership. Lilly Kolisko's report states:

> [...] This general meeting also had certain consequences for Count Polzer. On 29 April, he was informed that the Executive Council at the Goetheanum could no longer agree to him continuing to make the relevant communications to the members of the 1st Class. He was asked to return the text documents. Continuing his work would be considered unlawful.
>
> On 2 June 1935, Dr Fränkl read out a letter from the 3-person Executive Council at the general meeting of the AG in Austria, stating that Count Polzer was not entitled to hold these Lessions.
>
> A correspondence arose which also involved me in the matter, which is why I feel obliged to provide various clarifications here. It concerns a passage in a letter signed by Herr Steffen, Frau Marie Steiner and Dr Wachsmuth:
>
> "22 April 1935 [...]. At the Goetheanum - as long as Dr Steiner lived - there was no question of reading out his Class lessons at all; no one there received this assignment because no one expected Dr Steiner to pass away."
>
> Frau Kolisko only received it because - since she is so good at shorthand and could give the time to it - she was chosen by the Waldorf teaching staff to make the urgent request to Dr Steiner on behalf of the teaching staff: to allow her to travel from Stuttgart to Dornach for each Class Lesson, take shorthand notes, and read them aloud to the teachers who had joined the Class.
>
> This didn't even go smoothly, because when a teacher was later asked why he had withdrawn from participating in these lessons, even though he was a Class member, he replied: because it is untrue that the entire staff had made this suggestion and made this request. When the Stuttgart branch learned of this reading within this staff, to which,
>
> it seems, the teachers' wives were also allowed access, they asked Dr. Steiner that the reading take place on

Landhausstraße so that the other Class members could also participate..."

Regarding this letter, I can only refer to my previous statements, which explain exactly how the conveyancing for the teaching staff came about: *that it was a request addressed to me by Dr. Steiner. The letter is correct insofar as it states that it was not the teaching staff who approached Dr. Steiner with the urgent request.* Everything else is *a deliberate distortion of the true facts* by these three Executive Council members. It is an unforgivable act to try to dismiss this serious matter, which concerns the heart of the movement, by claiming that someone who can "take shorthand well" had been authorized by Dr. Steiner to do so. *What a disrespect for Dr. Steiner's personality this expresses.* A complete lack of understanding of esoteric facts and actions.[333]

In a letter to Ita Wegman dated 19 July, 1935, three months after the General Assembly resolutions, Lilly Kolisko addressed, among other things, her situation and the continued work of the Anthroposophical Working Groups in Stuttgart, co-led by Eugen Kolisko. Among other things, she stated:

Thank you very much for sending me your conference programmes. We also intend to organise a small conference in Stuttgart on Michaelmas for the Anthroposophical Working Groups in Germany.

Unfortunately, we have been given notice to vacate our beautiful premises on Archivstraße, so we don't yet know where we will be able to continue our work in the fall.

As far as my work is concerned, it is continuing quite well, except for ever-diminishing funds. Since the last General Assembly, I haven't received a penny from the Goetheanum. I think that's quite all right, but on the other hand the members of the "Association for the Promotion of the Biological Institute at the Goetheanum" should increase rapidly so that I can continue to work.

You will no doubt already have our new work programme in your hands and will see from it that we are trying to create something new for Germany. There is clearly a 5th working group on your programme (Dr. v.

Heydebrand). Unfortunately, we had to cancel this. Dr v. Heydebrand is leaving the Waldorf School, already at Easter and does not want to come to Stuttgart. We very much regret this, but we can certainly understand that it is too difficult for Dr v. Heydebrand to be here in Stuttgart without the Waldorf School and without her children. It is to be hoped that she will return after some time, when things have calmed down.

These working groups are a start and can be expanded as required if people are willing to join.

Yours sincerely, L. Kolisko[334]

The question also arose as to whether Lilly Kolisko could continue to read the Class lessons and admit members to the School of Spiritual Science, as she had previously received the necessary "Blue Cards" from Ita Wegman. Ita Wegman herself was determined to continue her Class work independently of the Goetheanum. In a letter to George Adams Kaufmann dated 19 June 1935, she wrote: "I have an inner commitment to this Michael School, and take responsibility for it." [335]

However, a letter from Ita Wegman dated August 10, 1935, reveals that she only had a few blue cards left for new members of the First Class of the School of Spiritual Science and was unable to obtain new ones due to her exclusion.

Dear Frau Kolisko! Some time ago, you sent me Miss Carita Stenbach's blue card. I have now signed it. It's better to do it this way than to give you a new card. I only have a small number of blue cards left so I don't want to issue new ones. I am therefore returning the signed card to you; I am sure you will be kind enough to give it back to her. Yours sincerely, I. Wegman [336]

Elisabeth Vreede wrote about this situation and about Lilly Kolisko to Willem Zeylmans van Emmichoven in The Hague in October 1935. As can be seen from Vreede's letter, Lilly Kolisko had great difficulty in distancing herself completely from the "blue card" system and indirectly from the relationship with the Goetheanum in matters of the First Class. She did not initiate or participate in autonomous efforts in this respect, although she was absolutely committed to the "working group" of Eugen Kolisko and his friends (such as Elisabeth Vreede and Willem Zeylmans):

[…] She [Lilly Kolisko] intended just to restart the Class with those who were giving the courses according to the new curriculum in Seestraße (the new location of the working group) but couldn't do it now. This was related to the second point: the blue cards.

It seemed to me that she had a general, more emotional antipathy to the idea that we "no longer want to recognise the blue cards", clearly seeing this as an abuse of our personal power over the Class. She also asked whether we thought we had the right to dispose of the Class in this way vis-à-vis the spiritual world. Of course, I was able to respond in the same way as in Clent [England], for many of her "aggravations" on this point were very similar to those of Dr Wegman. It appeared to me from what she said that she had not considered the full question of how things would have to go if the blue cards were to be retained. Here I was able to point out to her all the points that make our "line of behaviour" necessary. I think that this made an impression on her. I had the feeling that she suddenly realised that she hadn't thought things through practically as a whole. If she were a more agile soul, I think she would even have admitted this. But after a while, and also at the end of our conversation, she said typically: Yes, I can't go along with these decisions, which is why I can't hold a Class now. She said that nobody here would accept it if, after the green card, which had already been accepted with such difficulty, a "white" card was now forced upon them. I said that she didn't have to do that for Class lessons in a small circle, which was invited, so to speak, but then she spoke, not very clearly, about spreading out the participants. When I pointed to the new members to be admitted, she said that some should also be admitted. I think she imagined that she would pass these people on to Dr Wegman, who would then give them a blue card. My explanation that Dr Wegman had simply run out of them and whether she thought we should reprint them brought her to her senses a little, it seemed to me. I also said, and Dr Kolisko had told her the same thing, that each country could make somewhat different arrangements, but she didn't want to know anything about it […] Incidentally,

Frau K[olisko] has now become strongly involved in the
work of the working group again, she has, so to speak,
taken on the management of the new courses, feels
responsible for this work, so is no longer so lonely [...].
337

Due to the control of foreign correspondence in Germany, Lilly
Kolisko's contact with Ita Wegman in matters relating to Class lessons
and Class members became very difficult. In November 1935 the
Anthroposophical Society was officially banned in Germany – and all
connections with the Goetheanum were persecuted to the extreme. Lilly
Kolisko therefore wrote to Ita Wegman during a trip to Turkey in June
1936:

> Esteemed Dr Wegman!
> Since I am currently abroad, I would like to take this
> opportunity to address you on a matter concerning the
> First Class. Frau Frida Vollmar, a member from Buenos
> Aires, is currently in Germany. She has belonged to the
> Anthroposophical society since 1924 and would like to
> become a member of the 1 Cl. As she is a person with
> serious aspirations, I would like to support her request
> for membership. The lady will be coming to you soon.
> Perhaps you could issue her with a card. Mr and Frau
> Vollmar will be attending the conference in Bangor and
> will be travelling back to South America at the beginning
> of September. It would of course be very desirable if Frau
> Vollmar had the opportunity to listen to some Class
> lessons while she is in Europe. However, I don't know
> whether you will be doing anything in Arlesheim; it
> should not be later lessons, but just the early ones. But
> you can talk to Frau Vollmar about this yourself. The
> lady will refer to me and this letter.
> I could not write from Stuttgart as the letters are
> opened. I therefore do not expect a reply from you to
> Stuttgart. Perhaps I will see you in London in August.
> 338

Margarete Bockholt brought Ita Wegman's reply by letter with her to
England (to a conference in Bangor, see below):

Dear Frau Kolisko!

Frau Frieda Vollmar was once in Arlesheim with her husband. She asked me to join the Class. At the same time, however, she also said that she wouldn't be able to join any more Class lessons because she would soon be travelling to Argentina or Brazil.

Now I wonder whether it might be possible for you or Dr Vreede to give her the mantras and say something about them so that Frau Vollmar has something to work with. It would be very good if her husband could also be included. He seems to have a lot of understanding for anthroposophy, and since they are alone and no other members live near them, they could help each other if they both have the mantras.

What is the procedure for admitting members to the Class? Does membership have to be for 2 years or are two years of study also required? I would like to see the latter handled flexibly, because everything must be organised freely.

I am giving this letter to Dr Bockholt. I hope you have a good time in Bangor!

With warmest greetings, I. Wegman [339]

20.
Trip to Brussa (Turkey) with Elisabeth Vreede

Well, it was marvellous! The darkness, the mountain, the surroundings, the weather! We were on this mountain, only several 100 metres higher, at 2000 metres. Oh, how beautiful the corona is. I hadn't imagined it to be like this, so silvery, radiant, transparent and bright. It and Venus appeared all at once, Mars was not visible. This doomsday atmosphere beforehand and the joy afterwards![340]

The darkening that the sun experiences in the cosmos is also experienced by the gold solution on earth, and this is revealed in these experiences. [341]

In the summer of 1936, Lilly Kolisko travelled to Brussa in Turkey with Elisabeth Vreede. The common interest of the two very different women was to carry out experiments and observations during a total solar eclipse on 19 June 1936. The journey began on 12 June and took them through Yugoslavia, Bulgaria and Greece. On 15 June, they were in Istanbul and saw the Hagia Sofia, the archaeological museum, the Blue Mosque and a bazaar. On 17 June, the trip continued with a steamboat across the Marmara Sea onwards to Brussa.

The observations made by Lilly Kolisko and Elisabeth Vreede in different places and in different ways complement each other and enable a lively understanding of the total solar eclipse.

Lilly Kolisko transformed the hotel room into a laboratory and began her experiments. Elisabeth Vreede wrote in her travel letter to Arlesheim: "[...] Frau Kolisko has a retreat [...], and [...] Frau Kolisko has got wonderful pictures [...]."[342]

Lilly Kolisko stayed in her room the whole time and worked intensively. From 17 June, she carried out her experiments from 5 p.m. with "all substances" (with the 7 planetary metals); at the beginning every hour then every 10 minutes, and later, during the total eclipse on 19 June at 5.52 a.m., every minute with silver nitrate and gold chloride, until 6 o'clock; then again fading out once each hour until 20 June, 8 o'clock - without sleeping. She travelled immediately afterwards in the morning of 20 June back to Istanbul and in the evening on to Germany. This meant that her colleague Wilhelm Kaiser in Stuttgart was able to start work on the photographs on 23 June. Because of the rapidly occurring changes, especially in the pictures with silver nitrate, it was important to process the photos as quickly as possible.

Lilly Kolisko wrote in her publication "The Total eclipse of the Sun of 19. June 1936"[343] that her hotel room had a large balcony facing east, which gave her a good view of the sunrise and the eclipse. In this way, she was also able to perceive the external view and later notice the coherence with her capillary dynamolysis pictures. She described how, for example, the colours of gold chloride darkened during the eclipse and then lightened again. And how a darkness could also be experienced in the human soul, which only later gave way to a brightening again.

Elisabeth Vreede made her observations outside on the Uludag, a rocky plateau about 2500 metres above the Sea of Marmara, together with 100 other scientists and laymen.[344] Elisabeth Vreede's description of the solar eclipse was printed in the magazine "The Present Age" in England, edited by Walter Johannes Stein:

> In the early dawn we found ourselves together climbing up the last few hundred feet of the mountain-side, covered with juniper bushes and overgrown with violets. At the very moment when we reached the top the sun was rising over the mountain-tops to the east. There was as yet no sign of the eclipse; the sun was rising in the radiant light of a lovely summer's day [...]. Very few minutes afterwards the shadow of the Moon began to creep across the Sun. From the right-hand side the lunar disc began to encroach upon the Sun, eating its way ever more deeply in [...].
>
> [...] The visible surface of the Sun decreases, and the sky too grows darker, yet not in the same way as on the Earth. One feels a kind of estrangement between what is going on on Earth and in the Cosmos. Above the ineffable laws of cosmic rhythm are finding their fulfilment, bringing forth eclipses in their periods with marvellous regularity in space and time.
>
> Beneath is the Earth, robbed of the sunlight out of due

season, the victim as it were of an appalling misfortune, sick unto death and wretched-for such is the impression.

The feeling rapidly increases until the moment when totality sets in. Darkness is over the entire firmament; on the horizon only a narrow circle of light is still visible. Now to describe the last few seconds before totality: from the west we see approaching on the distant horizon a band of darkness, dirty reddish-brown in colour, extending rapidly towards us, though without well-marked outline. It is the actual shadow of the Moon, coming toward us with great rapidity from the west, to envelop the whole surrounding world in darkness for a few short moments. And at the very moment when the shadow reaches us (though the exact moment would be difficult to tell), in the Heavens to the east— that is, in the opposite direction-the miracle takes place: the light of the sun suddenly disappears, the black disc of the Moon entirely covers up the sun, and in the self-same second like a flash of lightning the solar Corona makes its appearance, and beside it the planet Venus. Both the Corona and Venus shine with a silvery light, seeming to overcome the impending darkness with their clear transparent radiance, reaching far around the blackened orb of the Sun. [...] So now the eclipse was complete. For a few precious seconds we were witnesses of the marvellous Corona, the only known solar phenomenon which is visible at no other time than at a total eclipse. [...].

[...] And now, no less suddenly than it had vanished, the light of the Sun returned. As in a lightning flash, the Corona, Venus, and the surrounding darkness dis-appeared. A minute point of brilliant light appearing on the left-hand edge of the Sun rapidly increased to a tiny disc of light, which for a moment seemed to be rotating about the left-hand edge of the Sun, and thereafter slowly increased in magnitude and intensity, repeating in inverse order the different degrees of partiality which we had previously witnessed. [...] But the remarkable thing was that from the very first moment the gradually returning light no longer had the uncanny quality which it had had immediately before totality. Feeble though it is to begin with, the diffused light that gradually extends over the

Earth feels normal and healthy. Gone is the nightmare of oppression that seemed to weigh upon the Earth. It is not only the gradual intensification of the sunlight which calls forth this impression. It is a radical difference in quality—an absolute polarity: formerly fear and anxiety, a sense of doom as of the end of the world; thereafter the brief and magnificent, no longer quite so uncanny interval of totality; and then at last the still feeble yet apparently normal sunlight growing from moment to moment. Nature appears to awaken from her frozen fear, the customary life unfolds again [...].[345]

21.
A new start in London

[...] My wife gives lectures here, and we are trying, steadily, to prepare the ground here for this biological work, which is currently still in Stuttgart [...][346.]

Unterlengenhardt and the Burghalde were not a permanent place for Eugen Kolisko; he encountered obstacles of various kinds, and the political situation became increasingly dangerous. In Lilly Kolisko's account: "After a relatively short time, however, he had to realise that it would not be possible for him to realise what he had set out to do as an ideal on the Burghalde. There was not enough understanding for it. In addition, he was increasingly convinced that the political situation in Germany would soon deteriorate to such an extent that all anthroposophical activity would be stopped altogether."[347]

Both Eugen Kolisko and Lilly Kolisko had been in contact with England for a long time. On 30 March 1936, Madeleine van Deventer wrote from London to Ita Wegman, who owned a clinic there (Kent Terrace, near Regent's Park and Steiner House):

> Kolisko was here for a few days. He has now decided to come here on 1 May, initially not as a doctor but as a speaker for the Society, to set up a kind of "training college" for anthroposophy, possibly together with Schubert and Heydebrandt [...]. Kolisko thus arrives on 1 May, initially without his wife, until he finds an activity or working opportunity for her [...].[348]

Two months later, on 7 May 1936, a side note in a letter from Ita Wegman to Fried Geuter, a curative teacher working in Clent, read:
> "[...] Mrs Merry, Miss Osmond, Mrs Frances are now only interested in Kolisko and Frau Kolisko, of whom [they] say that it is so important that she comes to England [...]."[349]

Lilly Kolisko then travelled to Bangor, Wales, in August 1936 for the Anthroposophical Summer School (Summer School on "Anthroposophy and the Free Life of Spirit in Normal College" [350]). She gave a lecture there with slides of her experiments during the solar eclipse in Brussa. During her stay she was able to visit the stone circles near Penmaenmawr, where Rudolf Steiner had also been during the "Summer School" of 1923.

In the autumn of 1936, Eugen Kolisko founded "The School of Spiritual Science and its Application in Art and Life" with the help of English friends, based on the title of the anthroposophical world conference eight years earlier.[351] Lilly Kolisko's name appeared twice in the school's first programme of events; she was to give public evening lectures at Rudolf Steiner House. On 11 November 1936 she spoke there about her experiments during the solar eclipse in June; a week later she spoke about the connection between planets and metals. Both lectures were accompanied by slides.[352]

Eleanor C. Merry, who was closely associated with D. N. Dunlop and Wegman, had helped organise the 1928 World Conference and was also involved in the new "School of Spiritual Science", owned a large flat in Regent's Park, very close to Wegman's Therapeutic Centre and Steiner House in Park Road. Eugen and Lilly Kolisko lived with her whenever they were in London during this early period, from 16 September 1936. Their daughter Geni stayed in Stuttgart until she had passed her A-levels - and then came to London.

On 18 November 1936, Eugen Kolisko wrote to Ita Wegman:

> [...] I now get together quite a lot with Dr Engel and Dr Neuhöfer [from the Kent Terrace Therapeutic Centre], as we have regular doctors' meetings, which are always well attended by new doctors. A wonderful man is the good Dr Mukerjee, who is now bringing [anthroposophical] medicine to India. I really enjoyed working with them and we have all become good friends. You have also met him. My wife gives lectures here, and we are trying, steadily, to prepare the ground here for this biological work, which is currently still in Stuttgart [...].[353]

Fig. 38: Lilly Kolisko in Penmaenmawr, August 1936.
From the estate of Cecil Reilly / S. T. Norway

Lilly Kolisko continued to look after members of the First Class of the School of Spiritual Science and repeatedly received enquiries from people who wanted to become members of the School while travelling in Germany. On 18 June 1937, she wrote to Ita Wegman and mentioned that she would work with individuals or small groups on the Class content during her visits to Stuttgart - Wegman was and remained her main point of contact in matters relating to the First Class.

> Dear Frau Dr Wegman! After a long stay in London, I will be travelling back to Stuttgart in a few days. One of the Stuttgart members, Mr Gravert, who was a member of the Anthroposophical Society for about 7 years, would like to become a member of the First Class. As you know, it is hardly possible to travel abroad; on the other hand, it is understandable that people are seeking access to the First Class despite all this. Nor can you write abroad on this occasion. I have now promised Mr Gravert to arrange his admission. My assumption that you would come here while I was in London did not materialise. I would have preferred to talk to you about this difficult general situation in Germany. Something has to be done for the people. In the meantime, I suggest that I send you the names from London and that you make a note of them, possibly with a number. After all, you can't send cards to the members. As soon as I am in Stuttgart, I will work with the individual people or in very small groups. I have known Mr Gravert myself for many years. He is a serious person with no objections to being part of the Class. I introduced him to all the difficulties that are currently arising both in Germany and as a result of the split in the Society in Dornach for every individual who seeks access to the Class today. Nevertheless, he would like to become a member of the Class through us. Please let me know in London whether you agree with this arrangement.
> Yours sincerely, L. Kolisko [354]

From June 1937, Lilly Kolisko regularly published articles on her research work in the English journal "The Modern Mystic"[355]. Her first article there contained a general description of the early days in Stuttgart with the first research tasks on foot-and-mouth disease, the spleen and questions that had arisen from the agricultural course.[356] She also wrote about the influence of the moon on plant growth,[357] the connection between metals and planets using capillary dynamolysis pictures[358] and about the "smallest entities".[359]

Marna Pease, the secretary of the Anthroposophical Agriculture Foundation, had already bought a small country house near London together with Eleanor C. Merry in 1928: "[...] With quite a lot of land, in which they can start a kind of market garden and also want to give Dr. Mirbt a place for experimental purposes, where the other farmers can then also meet and get advice [...]."[360] This "Old Mill House" was in Bray-on-Thames near Maidenhead, west of London, and so Lilly Kolisko was able to set up her laboratory there in a small bungalow adjoining the garden. A letter from Eugen Kolisko to Walter J. Stein shows that Lilly Kolisko moved the laboratory to Bray in September 1937. [361]

The new institute was named the "Biological Research Institute for Investigating Cosmic Influences on Earthly Substances and Living Organisms". Marna Pease and some of her friends wrote an appeal for support to buy the house and to cover the costs of remodelling it, the necessary laboratory equipment and other costs for the research work, including the salary of a laboratory assistant. [362]

On Saturday, 6 November 1937, a grand opening finally took place in the small bungalow. Eleanor Merry reported the following in "Modern Mystic":

> The "Modern Mystic" has published several articles by Mrs L. Kolisko dealing with her research and experiments in connection with the planets and the metals, the moon, and its influence upon plant growth, and which are the results of her studies in Rudolf Steiner's Spiritual Science. Readers will also have observed that an appeal was printed for funds to open a laboratory for this work in England. The "embryo" of this laboratory is now in existence. A small bungalow, adjoining the gardens at Bray, near Maidenhead, which represent the nucleus of the Anthroposophical Agricultural Foundation, has been purchased. It is intended later to build on to it an adequate laboratory where Mrs Kolisko will be able to carry out extensive experiments in various fields of this most fascinating form of biological science. On Saturday, 6 November, the bungalow, its modest rooms filled with the evidence of Mrs Kolisko's unique genius, was "opened" in the presence of a few friends. They represented the great numbers who have an unshakeable belief in the

immense practical value of the work which Mrs Kolisko has done and can do in the future.

I have known Mrs Kolisko for a number of years, and I look upon her always with astonishment. Never have I met any woman with so dauntless a faith and courage. Nothing is too hard for her to attempt; no expenditure of strength and time is too great for her. [...] By no means all of the results have yet been published; and we look forward to many publications in the future. A great part of her work is connected with medical as well as agricultural problems. But it goes without saying that the value and quantity of her future experiments must depend to a great extent upon the amount of financial help she may receive. [...] I know well that it has been no light task for her to leave her well-equipped laboratories abroad, and make a new beginning in a foreign land. I am sure that The Modern Mystic, with its ideal for bringing together Science and Mysticism, will wish her all success. [363]

Rudolf Hauschka also came to London frequently at Wegman's request to publicise his work at the Arlesheim clinic and to set up a laboratory for the production of remedies in Kent Terrace. Wegman hoped that he would do much to spread anthroposophy, anthroposophical medicine and curative education in the English-speaking world - he was worldly and professionally convincing. However, the old conflict with Lilly Kolisko had not resolved.

George Adams Kaufmann, who had always translated Rudolf Steiner in England and was closely associated with Wegman and Dunlop, wrote to Ita Wegman from London in June 1937, five months before the opening of the "Biological Research Institute" in Bray:

"[...] Wallace, together with Mrs Pease and others, wants to help bring about a laboratory for Frau Kolisko in England. The plan is to take a house for it in Bray; Wallace has co-signed the appeal for it. He has no idea of all the unfortunate differences, e.g. Frau Kolisko's hostility towards Hauschka etc., and is not the person to get involved. He is a person with a great deal of dedication and reverence and comes to us all, as he says again and again, as someone who wants to learn. I also very much hope that Hauschka, with his good knowledge of English, will be able to provide a great deal. Wallace, apart from being a doctor, is an enthusiastic natural scientist [...]."[364]

Ita Wegman's reply to George Adams shows that, despite her deep respect for Lilly Kolisko and her work, she had hoped for an understanding with Hauschka, wished for more mobility from Lilly Kolisko - and relied on Hauschka's commitment in London to Kent Terrace and anthroposophical medicine. This is how she wrote to Adams:

> As far as I know, there is not much chance that Frau Kolisko will realise that Dr Hauschka is also a meaningful person and can find many interesting things. It's so sad to always have these stupid stories between people [...]."[365]

Eugen Kolisko subsequently refused to allow Rudolf Hauschka to give a course at the School of Spiritual Science, and the "stupid stories between people" continued. On 4 July 1936, however, Adams wrote in another letter to Wegman:

> "[...] Later, Stein happened to come in and heard about the incident. He said that Kolisko was in a difficult situation because of his wife; he himself (Stein) had tried everything to bring Frau Kolisko to her senses about Hauschka, but in vain. I finally decided not to push the matter any further, especially as I know from Hauschka himself that he doesn't want it. Kolisko then also apologised, and since then (it was about 10 days ago) we have rearranged some things, he himself is visibly pulling himself together [...]."[366]

Meanwhile, Geni Kolisko had finished school and wanted to study eurythmy in Dornach, despite all that had been inflicted on her father (and to some extent her mother) there. She wrote a letter of application from London on 18 November 1937, as can be seen from a note that is now kept in the Rudolf Steiner Archive: "From London, 18 Nov. 1937, Frau de Jaager gives this little letter from Genie Kolisko to Frau Doctor [Steiner] to read. If the doctor agreed, Frau Jaager might write her a polite letter saying that it would not be possible to take her in without her parents asking and taking care of her here - or at least that she would be of legal age. Or: The parents would at least have to give their consent to take care of her here when she attends school - if they don't want to enquire themselves. Besides, she was still very young (about 18 years old), so she could wait a little longer."[367] It is not known what Geni's letter said and what the reply from Dornach was; the fact is, however, that she went on to study eurythmy there.

22.

Journey to India, 1937

In India, Frau Kolisko obtained particularly good rising pictures with the metals of the upper solar planets. On the sea, she usually got good pictures with mercury salts (e.g. Hg+ Ag, also Ag+ Hg+ Fe). When she travelled through the Suez Canal, there was a solar eclipse, which could be followed well from the known pictures.[368]

Dr Mukerjee invited a large number of these doctors to his house, and Kaviraj Jyotish Chandra Sen read an introductory report on the subject. He brought with him, for purposes of demonstration, a considerable quantity of remedies, principally metallic, in different stages of preparation. After his explanations were concluded, I spoke briefly about my experiments made at the Biological Institute in connection with planets and metals. It was a most interesting meeting! Ancient Indian medicine and scientific investigations found themselves united on a common ground.[369]

Lilly Kolisko was invited to India at the beginning of 1937 by the Maharani of Travancore (the wife of the Maharaja of Travancore) to carry out experiments; this contact had apparently come about through the Indian homeopath Amarnath Mukerjee (Calcutta), about whom Eugen Kolisko reported to Ita Wegman in his letter of 18 November 1936 and who had also met Ita Wegman in London ("A wonderful man is the good Dr Mukerjee, who will now bring [anthroposophical] medicine to India. I have had a lovely time with him and we have all become good friends. You also met him.")

Lilly Kolisko had planned to travel as early as April 1937, but the journey only became possible at the end of the year thanks to invitations from various people. Walter J. Stein, who had emigrated to England with Ita Wegman's help, wrote to Wegman in May 1937:

[...] Frau Kolisko did not travel to India but postponed her ticket to October. The Maharaja's wife, Maharani von Travancore, who invited her, is currently in Java. Besides, the hot season had already arrived, and Veltheim and Mrs Pot advised against it, as did Dr Mukerjee. Only Mrs Dodwell wanted her to come. And to come now. It was decided to postpone it after all. All things can be better prepared for October [...].[370]

Lilly Kolisko's journey took her by ship through the Suez Canal; she already carried out her experiments with capillary dynamolysis en route, as can be read in the minutes of the Stuttgart Medical Circle, to which she reported on her research and her journey after the Second World War:

In India, Frau Kolisko obtained particularly good rising pictures with the metals of the upper solar planets. On the sea, she got exceptionally good pictures with mercury salts (e.g. $Hg + Ag$, also $Ag + Hg + Fe$). When she travelled through the Suez Canal, there was a solar eclipse, which could be followed well from the known pictures.[371]

Walter Johannes Stein wrote in the magazine "The Present Age" in February 1938 about the course of Lilly Kolisko's journey:

Her tour is extensive, starting from Cochin she travelled to Madras and first made a trip around the province where she was invited by the Maharaja of Travancore. Travancore is the southernmost province of India and is remarkable for its beautiful scenery and pure Indian character. Here she gave numerous lectures. She then travelled on to Calcutta, Benares, Delhi and Darjeeling etc., and will later visit Madras, Bombay and the Dutch East Indies, where she was also invited to give lectures.

In Calcutta, she took part in the scientific congress of the British Association, which brought together many English and Indian scientists.

There she also met Baron von Veltheim - an old friend (papers from his Indian travels have been published in this journal). Mrs Kolisko was invited by Dr A. N. Mukerjee, the president of the Homeopathic Society in

India, to give lectures on the medical and agricultural aspects of her work on metals and planets, the moon and plant growth and on the "Efficacy of the smallest entities". These lectures, as we can read from her letters, aroused great interest, especially because this knowledge is not unknown to students of ancient Indian scriptures.

Mrs Kolisko also visited Rabindranath Tagore in his house in Santiniketan, where she was warmly welcomed by the old philosopher, who had just recovered from his illness. Here is a travel report from Mrs Kolisko:

From Calcutta to Darjeeling

"[...] The railway winds in serpentine fashion from 1,000 to 6,000 feet. This of course takes a long time. The surroundings are enchanting. Small hills - becoming mountains - covered with terraces of tea-plants. I am astonished, seeing how cold it is, that tea is able to grow.

The vegetation gradually takes on an alpine character: innumerable ferns and bracken, like forests--trees resembling Araucaria, but which have branches like weeping willows. The people are quite different from the Bengalese: yellow faces, prominent cheek-bones, narrow eyes sometimes quite tiny. Both men and women are small in stature; the men wear long pig-tails and at the end of them a big tassel; the women have rings in their noses, upper lips, and ears - silver bracelets and anklets. They are very friendly people. The children wave to us. There seem to be many blind people among them.

On arrival in Darjeeling, I could not see the highest of the mountains as they were hidden in mist. In Calcutta they had told me that in the rainy season it was nearly impossible ever to see anything of the high mountains. There are lovely things in the Hotel, Tibetan ornaments, prayer rolls, bronze, copper and silver plates, and Buddhas of bronze.

The next morning I went out before sunrise. It was very cold and I needed my furs. I had no guide and it was not easy to find my way in the darkness. But I managed to climb up some distance. The sun rose in a crimson glow. Opposite, the crags of the Himalayas shone rosy-white.

Curiously enough, this time I had no desire to climb the
heights. It seemed to me much more desirable to stand and
gaze at them. If one could ever plant one's feet on their
summits one would feel oneself a conqueror - but at the
same time would cease to feel the same wonder at them
and appear to oneself greater than one really is.

No sooner had the sun risen than the mists below began
to heave and move and gradually to ascend. I feared they
would hide the mountains; but it remained wonderfully
clear above.

When I got back to the Hotel I could see the mountains
directly from my window. They gleamed brilliantly white,
- while in my room the fire crackled. If I could stay here
longer, I could make many experiments [...]." [372]

Another travel report appeared in the May 1938 issue of "The Present
Age" - this time entirely by Lilly Kolisko, with a report from Madras
dated 14 February 1938:

Preliminary report on my stay in Calcutta: I arrived in
Calcutta on the 2 January 1938. During my stay there I
was the guest of our Indian friend Dr A. N. Mukerjee,
the leading homoeopathic doctor in Calcutta. He saw to
it that my sojourn in Calcutta was made so interesting in
every way that I shall have to reserve a fuller description
of it for a later occasion.

In India not only the usual modern allopathic and
homeopathic systems of medicine are practised, but also
the so-called Ayurvedic medicine, which has its source
in ancient Indian wisdom. Dr Mukerjee made it possible
for me to meet some Ayurvedic doctors - *Kaviraj* - who
willingly gave me information about this most
interesting domain of old ancient Indian knowledge.

Dr Mukerjee invited a large number of these doctors to
his house, and Kaviraj Jyotish Chandra Sen read an
introductory report on the subject. He brought with him,
for purposes of demonstration, a considerable quantity of
remedies, principally metallic, in different stages of
preparation. After his explanations were concluded, I
spoke briefly about my experiments made at the
Biological Institute in connection with planets and metals.
It was a most interesting meeting! Ancient Indian
medicine and scientific investigations found themselves

united on a common ground.

Kaviraj Chandra Sen gave me the manuscript of his lecture for publication in the "Present Age" and promised to contribute further articles. The present article will, to begin with, be printed without the addition of any comments from me. I propose to discuss various details of it later. In the meantime, I would like here to express my most sincere thanks to Dr Mukerjee's friendliness he showed me in Calcutta and for all the trouble he took to make my visit interesting and fruitful. I am also most grateful to Kaviraj Chandra Sen for his paper on Ayurvedic medicine and hope that in the future we may find a way to establish useful co-operation.[373]

Fig. 39: Lilly Kolisko with the homeopathic doctors in Calcutta, 1938.
From the estate of Cecil Reilly / S. T. Norway

She also reported on India and the discussions she had there in her book "The Agriculture of Tomorrow":

During my travel in India, I tried to find out as much as I could about old customs, or still practised customs connected with the moon. Wherever I asked, people began to talk about the new moon, never about the full moon. There are many rules about the new moon. The new moon day is even an official holiday. Nobody would undertake a journey or do business when the moon is new. One does not even call a doctor, because the cure could not be successful if the doctor starts the treatment on a new moon day. There is one special new moon in the year, when no child may be born. Should this happen, everybody is convinced that the child will have a bad character and may become a thief or even worse. It is interesting then to find out that there is also a special month during which no marriage can take place; and if this rule is observed, then there is no possibility that a child will be born at that unlucky new moon.

During a visit to a hospital in Madras I had an interesting talk with the English doctors residing there. I told them about my experiments concerning the influence of the moon and asked if they were able to tell me something about the native customs there. At first, they did not remember anything, but later they told me the following story: A woman patient was completely cured, and the time was fixed for the husband to come and fetch her. The husband came, but he declared he would not take his wife home that day. She could only leave the hospital the next day. Why? It was new moon day, and if the wife left the hospital when the moon was new, she would fall ill again.

Furthermore, the doctors told me, that the Indians believe, that, if, during the critical stage of an illness, a new moon day occurs, then the patient will inevitably die. And, strange to say, the doctors added, that this really had happened several times.

In Travancore (southern India) I once asked an Indian who began to speak about the new moon, why I never heard anything about the full moon. He smiled and said: "You see, the full moon is always good, we need not speak about it. But you must be careful about the new moon".

Of course, the modern, educated Indian, who is very proud of his English degree, does not like to speak any more about the influences of the moon. Western Science has taught him that these old traditions are not true, not scientific. I shall never forget an interview I had with an Indian professor at Madras University. I gave several lectures about the "Influence of the Moon on Plant Growth" in Madras; and, talking with this professor, I felt that he grew more and more uneasy. "Your experiments are really very interesting, and, of course, I believe you are right; but you see, I find them dangerous".

That was the first time I had ever heard that my experiments were "dangerous". Why? Because the Indians are just beginning to learn to forget their old traditions. They want to be *scientific*! And if it is true, that the moon has an influence on plant growth, then people will say: "then all our other traditions are also true". And that is not possible. There are so many things which cannot be true.

It is precisely one of the tasks of a true scientist *to find out what is true, and when he has found it, to say so, even if it does not please people...*[374]

She published a further description of her encounters in India in the book "Physiological and Physical Proof of the Efficacy of the Smallest Entities 1923 - 1959":

During my stay in India in 1938, I had the opportunity to learn about the Indian attitude toward homeopathic medicine. I owe this to my Indian friend Dr. Amarnath Mukherji in Calcutta, who also took me to various factories that produced homeopathic remedies. It was interesting to see the enthusiasm and dedication with which the remedies were prepared strictly according to Hahnemann's instructions. In one factory, I watched a man prepare the potencies in relatively small bottles, vigorously hitting the bottle against the ball of his thumb, waiting a while until the liquid had settled again, and then repeating the process as often as Hahnemann prescribed. I was bombarded with many questions about my research and was just as willing to answer the questions I asked. [375]

I was surprised that I only ever saw the relatively small bottles, containing about 100 cc of liquid, in use and asked what would be done if a larger quantity of a remedy had to be prepared. It would be difficult to potentise a larger quantity in such a way that the bottle would hit the ball of the thumb. I was told with a smile that the potencies would then be prepared in small bottles as often as necessary until the required quantity was reached. But that must take a lot of time, I said. "Oh, the time, that's not a problem for us," was the smiling reply.

In one of the largest pharmaceutical companies, at M. Bhattacharyya & Co. in Calcutta, I was also willingly shown the preparation of potencies, trituration's, the production of granules, tablets, etc., and at my request, I received several mother tinctures of Indian medicinal plants for experimental purposes. I wanted to potentise these substances after my return to England and conduct growth experiments with them to see to what extent they were comparable to my other experiments [...].[376]

Lilly Kolisko's journey lasted a total of six months; after a long voyage by ship, she reached London again at the beginning of June 1938. On 1 June, Eugen Kolisko wrote in a letter to Ita Wegman:

"[...] My wife is back from India [...]."[377]

23.
Outbreak of war. Eugen Kolisko's death, 1939

[...] The declaration of war had a devastating effect on the number of participants in the Rudolf Steiner School (School of Spiritual Science). All the young people were drafted or volunteered for military service. [...] At night we looked up at the starry sky and he [Eugen Kolisko] talked about the interesting constellations, about Jupiter, Saturn, Mars and the imminent Great Conjunction. Soon, soon he would see everything, gain a real overview, then he would begin to write the various books that seemed necessary to him [...].[378]

[...] Quite apart from the heavy loss that this death means to us, I am shocked by the grace and greatness of such a death! I clearly feel how his liberated being breathes in those immense expanses, free from all petty earthly ties - those expanses that his spirit had always striven for here, for which he struggled [...].[379]

At Whitsun 1938, a youth conference was organised at Wynstones Waldorf School in Brookthorpe, Gloucester. It was the result of weekly discussion evenings with young people in York Gate, where the Koliskos then lived.

Eugen Kolisko was one of the speakers. It is uncertain whether Lilly Kolisko - who had just returned from her trip to India - also took part. The other major anthroposophical conference that year was the "Summer School" in Bangor, to which several friends from Central Europe also travelled, including Ita Wegman.

The situation in Europe following the annexation of Austria to the German Reich in March 1938 brought many refugees to England. The Koliskos helped fleeing anthroposophical friends of Jewish origin, including Karl König, who arrived in London on 8 December 1938 and visited the Koliskos the evening after his arrival. [380]

After Lilly Kolisko's journey "to the east" in 1938, Eugen Kolisko travelled in the opposite direction at the beginning of 1939, "to the West", to America, "to scout the ground for anthroposophy". Lilly Kolisko wrote:

He travelled to America, from which he returned full of hope. He had again met many new people, established many new relationships and expressed the hope: "If God gives me three more years of life, I hope to break through." I was puzzled by this strange expression. Why should it only be three years? There was still - so I thought - a long time ahead of him.[381]

After Eugen Kolisko's trip to America, from 19 July 1939 the Koliskos lived in their own flat in London, St. John's Wood (City of Westminster), also not too far from Regent's Park, Rudolf Steiner House and Ita Wegman's Kent Terrace clinic.

Soon afterwards, on 3 September 1939, Germany declared war on England, which changed Koliskos life in London:

"We were foreigners, and our movements were limited to a small radius. All foreigners had to appear before a tribunal that more or less decided their fate. We were both called before the tribunal and treated with great courtesy. We answered truthfully to all the questions put to us, and the chairman of the tribunal then declared us 'friendly aliens'. We were allowed to continue working in England [...]."[382]

On 27 September 1939, Vera Piper from the Kent Terrace Clinic informed Ita Wegman:

The situation here is so bad right now that I couldn't do what you suggested with the house. Everyone is leaving London as quickly as possible – thousands of schoolchildren have already left. We have no patients, everyone is calling and saying they won't be coming now until the war is over. We all received gas masks today – our beautiful Regent's Park is full of trenches against the air raids [...].[383]

Lilly Kolisko later wrote, looking back on this time and the completely unexpected death of Eugen Kolisko from a heart attack on 9 November 1939:

The declaration of war had a devastating effect on the number of participants at the Rudolf Steiner School (School of Spiritual Science). All the young people were drafted or volunteered for military service.

November 1939 arrived – it was often damp and foggy. Dr. Kolisko felt that November was the worst month of the entire year. When it was over, things would get better again. Until December arrived... At night, we looked up at the starry sky, and he talked about the interesting constellations, about Jupiter, Saturn, Mars, and the upcoming Great Conjunction. Soon, soon, he would survey everything, gain a true overview, and then he would begin writing the various books he felt were still needed. A book about the school doctor, about chemistry, the study of human beings, etc. He still seemed to be looking confidently toward the future. Yes, one could say he was a person who lived more in the future than in the present.

It was on the morning of November 26th when he came to breakfast, strangely excited yet joyful, and said: "Just imagine, today I dreamt about Dr. Steiner. I walked through a long, dark corridor, and when I finally came out, Dr. Steiner was standing before me, holding out his hands. I was so full of joy that I simply threw my arms around his neck. [...]."

On November 28th, or early in the morning of November 29th, he came to my desk, threw a pack of notebooks in front of me, and said: "Now I know everything; tomorrow we can start writing the book on agriculture." I replied doubtfully: "Do you really know everything?" "Yes," he replied, "now everything is perfectly clear to me, and we can write."

On November 29, we both wanted to go to Bray, where the Biological Institute was located. But phone calls kept coming, holding us back. He had to quickly see a patient but promised to be back in time. He came, and we made our way to the bus. Again, a lady stepped in our way and wanted to quickly get a prescription from him. He did that, too, but time was a bit tight, so he suggested taking a taxi. "I won't kill myself for two shillings..." What a strange expression, it flashed through my mind: "I don't want to kill myself for two shillings."

We arrived at Paddington Station, when he remembered that he needed to quickly call Dr. Engel. I waited in front of the telephone kiosk, keeping an eye on the station clock, as our train's departure time was approaching. Then I knocked quietly on the door to alert him. We quickly walked along the platform. Again, he stopped because he had to quickly buy a newspaper, as he thought it might contain something about Dr. Zeylmans. I watched him reach into his pocket to take out the money and continued walking, as he would soon catch up with me. I lost sight of him in the crush, but I was sure he was behind me. The train was close to departure; I looked along the platform but couldn't see him. The departure signal sounded, and I boarded, thinking he had probably already boarded a compartment. Arriving in Maidenhead, I waited for all the passengers; he wasn't among them. What should I do? He must have missed the train and would be coming on the next one. So, I went to Bray and waited.

He didn't come on the next train either. In the afternoon, I asked Mrs Pease to call London but couldn't get a reply from there either. So, contrary to my original intention. I went back to London.

When I got home, Dr. Stein approached me and said, "I wanted to tell you that Eugen had an accident." Then he stopped: "Oh, there's no point, I have to tell you that Eugen is dead." I remember replying, "That's not possible..." "Yes, yes, he's in the hospital, and we can go right away. The police sent a police officer to wait for you. But I wanted to spare you the shock, so I came myself to tell you."

Dr Kolisko had collapsed at the station, was taken to the waiting room, pulled himself together again and said he had to catch the train because his wife was waiting for him. He did catch a train, but not the one I had left on, and was later found dead on a seat there. These are the bare facts.

Dr Stein, as a loyal friend, helped me through the first few days with the necessary preparations, supported by his wife. Other friends also helped as much as they could, and I owe them my thanks for their selfless help.[384]

Magda Meyer lived as a child in Rudge Hill House in Edge, near Stroud, not far from the Wynstones Waldorf School, and told how Arno Rohde broke the news of

Eugen Kolisko's death to the children: "The man who loved all the children has died."[385]

Willem Zeylmans van Emmichoven travelled to England on 27 November. He was invited to a medical consultation with Wyneberg in London and was able to visit Lilly Kolisko during these difficult days.[386]

The cremation then took place on 4 December 1939, in Golders Green, with the funeral rites of the Christian Community. The difficult war situation meant that Eugen and Lilly Kolisko's daughter, Geni, arrived late for the cremation and was unable to return to her eurythmy studies in Switzerland afterward, as England's borders were closed due to the war. Lilly Kolisko had turned 50 at the end of August 1939, and Geni was 20 when Eugen Kolisko died at the age of 46.

The news of his death reached his friends in Central Europe only slowly because of the war.

On 9 December, Maria Röschl wrote to Ita Wegman from Wynstones, where she worked at the Waldorf School, on the day of the funeral service: "Quite apart from the heavy loss that this death means to us, I am shocked by the grace and greatness of such a death! I can clearly feel how his liberated being breathes those immense expanses, free from all petty earthly ties - those expanses that his spirit had always striven for here, that he struggled for ..." [387] On 11 December 1939, Walter Johannes Stein, probably Eugen Kolisko's closest friend and companion, wrote a detailed letter to Ita Wegman:

> Dear Dr Wegman,
> As this letter probably will reach you at Christmas time I wish to convey to you and to the friends in Casa Andrea Cristoforo my best wishes for the festival and the coming future.
> You know that we have been going through a most embarrassing time in losing our friend Dr Eugen Kolisko in a most sudden and unexpected way. His death was caused by thrombosis of the coronar arteria. It is difficult to say how this has happened. It was either due to a previous fall, which has caused a concussion from which the clot may have originated, or it was due to an earlier illness which has not been recognised. Frau Kolisko thinks to remember now that he was breazing a little more with difficulties than usually and so on, but this is all very vague, as he was not really ill, but was

taken from us in midst of an immense activity. He had just assisted in founding a second journal "Tomorrow", which promised to be a success, and he had started to write books. It happened at Paddington Station on November 29th about 12,30 to 12,45 p.m. He went to the station together with Frau Kolisko to travel from there to Bray. But immediately before entering the train he left her in order to buy a paper, the Daily Mail, because he had heard that this paper did contain a notice about Dr. Zeylmans having been called to London for medical consultation. So, he went to buy this paper; but. Kolisko did not see him coming back, she went back to look after him, but could not see him. So, she explained that to herself saying, "He might have taken the subway and is already in the train." But when she came there, he was not in the train. So, she went on to Bray alone waiting in Bray for the next train, but he did not arrive. Then she started to phone to friends.

In the meantime, Kolisko really had taken the subway running upstairs to catch the train. He collapsed and was taken to the first aid's post. Here he soon recovered drank a glass of water and went away to find his wife, and when he did not find her, he took the next train to Bray. But in this train, he collapsed again and was taken out before the train left, already dead. Then his body was taken to St. Mary's hospital. It took ten hours before we learnt what had happened, and it was my task to tell Frau Kolisko when she returned in the evening, and to accompany her to the mortuary of the hospital. It was very tragic as Frau Kolisko was not present when it happened and only heard of it ten hours later when I got the message by phone. I went to her home, but she was not there. It was also impossible to reach her in Bray, because she was in the train back to London. In the meantime the police appeared to tell her. The policeman was very kind and nice and left the papers to me asking me to convey the message to her as it would be a too great chock if she should receive the message through him. So, I waited in the gangway until she opened the door, still knowing nothing. It was a terrible experience, and nobody would have ever thought that this life would end like that.

Life has become difficult, and we struggle along as good as we can. Few days ago, we have been sitting together with Frau Kolisko, sitting around the fire place and reading out letters he has written. I have preserved all his letters since 1904 as we went already together to school. And so we were living in memories. Geni Kolisko, his daughter, has in the meantime arrived from Switzerland, some hours too late to attend the cremation.

Now, as a few days have passed we have found back to our daily work and to the attempt to continue our tasks [...].[388]

One day later, on 12. December 1939, and not yet in possession of Stein's letter, Ita Wegman turned to Lilly Kolisko:

Dear Frau Kolisko,

I was deeply stirred by the news of Dr Kolisko's death and now as I gradually hear from one end another how it took place I must write to you.

Such a death is in keeping with his great personality - so tragic - as was his whole life - that of a great individuality who did not find the place to live on earth as he needed. Once Rudolf Steiner said to me that his life now was a preparation for the next, so he will not have long to wait for that one in which he can fully develope. So we think of him, dear Frau Kolisko, who has gone before preparing the way for us. I do hope the shock has not been too hard for you.

It is good that Geni is with you. I hear her first words were, when she heard the news, to go as quickly as possible to London to be with you.

With my deep sympathy and greetings, Yours sincerely Ita Wegman.[389]

In March and April 1940, Ita Wegman wrote a remarkable obituary for Eugen Kolisko at the Casa Andrea Cristoforo in Ascona, where she was staying at the time. [390]

Many years later, in the biography of his life, Lilly Kolisko described a spiritual experience after Eugen Kolisko's death:

After death: We know that it is possible to accompany souls for a while on their journey through the gate of death. You can learn a lot, experience a lot, from souls close to us who have departed. Dr Steiner has often described the ways to do this. These human souls can impart important things to us, but - what they tell us is to be treated as a great gift and not intended for sharing with others. Great mistakes can be made in this respect if we do not heed the instructions we receive. Those who do not know how to remain silent lose contact. However, it is also possible that the deceased person close to us wishes that a certain message be passed on to his friends or other people close to him. After some time had passed and the soul looked back on its past life on earth, it went through certain experiences. It became particularly clear to her that she had made a certain mistake: she had tried too early to realise physically what she had spiritually recognised as right. She wanted to share this realisation with others. Dr Steiner himself told us that Dr Kolisko's death should be seen as a sacrificial death. This brought much warmth into the spiritual worlds, a rekindling of the Christmas conference impulses ... [391]

Lilly Kolisko kept Eugen Kolisko's urn until she herself died in 1976 and the ashes of both were scattered together in the earth.

Fig. 40: Eugen Kolisko. S. T. Norway

24.

Finding a new home - Edge

Dear Dr Wegman! Thank you very much for your letter. Following the death of Dr Kolisko, Frau Kolisko was unable to maintain her flat in London and has accepted an invitation to stay with friends in the country. She is moving these days. She hopes to have time to finish the book there which she worked on with Dr Kolisko, which is well advanced [...].[392]

[...] It is my duty to write the book now with the help of his notes and remembering our talks upon all the various aspects of the subject. But it will still be our book, the fruit of our common studies and the work of many years together. May it help to solve the urgent problems of our present time.[393]

Lilly Kolisko tried to continue her life in London, but she found it difficult - also from an economic point of view. She earned a small income from her articles in various magazines. In 1939 she wrote monthly in "The Modern Mystic" under the title "Astro-Biological Calendar". The last article in this series appeared in January 1940. She reported on her experiences in the annual cycle with capillary dynamolysis, on germination experiments underground and in the laboratory and on the correct sowing times during the different phases of the moon. She always added a table for the coming month with moon phases and planetary constellations in these articles.[394] There was also a series of articles about her experiments in another magazine, "Tomorrow", at the beginning of 1940 with metals and planets. [395]

In March 1940, Lilly Kolisko took part in a curative education conference at Rudge Hill House in Edge. Arno Rohde had arranged the conference - the occasion was the inauguration of a house for a curative education children's home. Lilly Kolisko's lecture topic was "Agri-culture of Tomorrow" - just like the title of the book she was co-authoring with Eugen Kolisko and which she now wrote alone.

Fig. 41: Edge. Private property

Fig. 42: Rudge Hill Cottage, Edge, Stroud. Private property

The anthroposophical doctor of Jewish origin Norbert Glas, who had run the sanatoriums on Ita Wegman's behalf until the invasion of the German troops and the appearance of the SA in Gnadenwald,[396] also gave several lectures during this conference. He and his wife had emigrated to England with Wegman's help and lived for a time with Arno Rohde at Rudge Hill House. Martha Meyer from Stuttgart, who was on friendly terms with Lilly Kolisko, also lived there with her children Magda and Georg Meyer during the war years.

Lilly Kolisko was then offered, probably by Arno Rohde, to move into a small gardener's cottage nearby: The Rudge Hill Cottage. The cottage was very primitive, had neither electricity nor running water, but windows in every direction, which was particularly important for Lilly Kolisko: she was able to set up a new laboratory on the first floor. This house was her new home from April 1940 until the end of her life in November 1976, i.e. for 36 years.

She now received many enquiries about Eugen Kolisko's estate. Ita Wegman also wrote to Walter Johannes Stein on 31 March 1940:

> [...] Dr. Husemann asks, if he could have the notes, which Kolisko made on the work with the children at the Waldorf school, he would, together with his own experiences, like to write a medical book. He asked me if this would be agreeable to Frau Kolisko, and so I ask you if you think this is possible? Perhaps you would be so friendly and inform yourself on this matter? I do not know whether it would be agreeable to Frau Kolisko if I were to put the question [...].[397]

Stein replied on 8 April 1940:

> Dear Dr Wegman! Thank you very much for your letter. Following the death of Dr Kolisko, Frau Kolisko is unable to maintain her flat in London and has accepted an invitation to live with friends in the country. She is moving these days. She hopes to find time to finish the book there which she worked on with Dr Kolisko, which is well advanced.
>
> From conversations with her, I know that she intends to work out everything else that is more or less sketchy herself; and that she has rejected all requests to transfer any material to other people for this purpose. She sees her further task precisely in the elaboration of these

things. It would therefore lead to no result if I were to pass on your enquiry. She is not free from bitterness after the many experiences she has gone through, and I would only provoke more of it without achieving anything. I therefore ask you to understand if I cannot help you in this matter. She would probably not even be able to understand that you are not even in this for yourself or your co-workers, but for Dr Husemann. I also find that difficult to understand. I remember that Dr Husemann worked with Dr Zbinden, so I can't understand at all how you are interested in this matter. But I have been away for so long now and am so disorientated about everything that has happened in the meantime that I may well lack the prerequisites [...].[398]

In fact, Walter Johannes Stein was somewhat "unorientated" in this respect - Ita Wegman had arranged the request from Gisbert Husemann, the Stuttgart school doctor and successor to Eugen Kolisko in the care of the children, but not from Friedrich Husemann (Sanatorium Wiesneck), who had taken over the medical section leadership in Dornach (together with Walter Bopp, Richard Schubert and Hans W. Zbinden) after Ita Wegman's dismissal in 1935 - and was no friend of hers.

Meanwhile the war continued - on 12 November 1940, Maria Schmidler wrote to Lilly Kolisko from London:

> [...] Bombs fell in the High Street, Arther had an incendiary bomb, Francis moved out, 4 bombs came so close to them in one night! At the Old Mill House, 2 bombs fell so close that all the windows were smashed. It's interesting how many people have a "narrow escape", the angels are so diligent [...].[399]

Lilly Kolisko also had a narrow escape; she had rescued her laboratory from the Old Mill House in Bray to Edge in the nick of time.

Agriculture of Tomorrow

Norbert Glas informed Ita Wegman on 12 April 1941:

[...] Frau Kolisko, who is living two minutes from our house, is very busy. She has just finished a new and very interesting book about "Agriculture of Tomorrow". There are 16 chapters with many pictures of her experiments through many years. The main question is also: where is the editor? But I think there is some hope he will be found. [...][400]

On 21 March 1941, the book was completed, but the war and the lack of paper, among other things, prevented the publication of the book with no less than 424 pages and 299 illustrations. [401]

It was clear that the research described in this book was an intensive collaboration between Lilly and Eugen Kolisko, especially in the medical field. Shortly after Eugen Kolisko's death, at Christmas 1939, she wrote a preliminary note to the book, which states:

It was originally intended that this book should be written in two parts, by Dr. Kolisko and myself working in collaboration. He had collected his material during many years. In the early morning hours of 29 November 1939, he called me and said he was ready to begin. He was full of enthusiasm and energy for the task. A few hours later he suddenly passed away.

Yes, we were both ready. It is my duty to write the book now with the help of his notes and remembering our talks upon all the various aspects of the subject. But it will still be our book, the fruit of our common studies and the work of many years together. May it help to solve the urgent problems of our present time.[402]

In this way, the two of them took a life review together: Eugen Kolisko in his own way on the other side of the threshold and Lilly Kolisko through the processing of their collaboration since 1920.

Ilse Ketzel, who worked at Wynstones Waldorf School at the time, noted in her diary:

29 November 1942

It is the third anniversary of Dr. Kolisko's death day. Mrs Kolisko was able to bring Dr. Kolisko's library to Wynstones School near Gloucester. The archive, once upon a time a class-room for chemistry became a most interesting room with Dr Kolisko's furniture from Stuttgart, many books, pictures, crystals. Mrs Kolisko is reading lectures by Rudolf Steiner there. Dr. Stein is giving every Friday-night a public lecture, which all the teachers of the school and several other people attain. Today the room has a special festival appearance and Mrs Kolisko has made some beautiful flowers out of coloured wood-fibre. So, the memory of Dr. Kolisko is living on [...].[403]

A grandson is born

A significant event for Lilly Kolisko in these dark times was the birth of her grandchild at the end of 1942. Her daughter Geni taught eurythmy at the Wynstones Waldorf School,[404] where many refugees had been accommodated: Geni met a young man from a Jewish family there. However, the relationship was not accepted by this family, and when she had a child, she became a single mother. Walter Johannes Stein invited her to stay with his family in London so that she could be there during the birth and the early days with the baby. But because London was bombed so frequently, she soon moved to the safety of Edge.

For Lilly Kolisko, this little grandson was a great joy during the difficult years. She called him "my eye stone". And later, when Geni and he no longer lived with her in Edge, she visited them both in Wynstones, and it was very important to her that her grandson heard many fairy tales - which she told very vividly during her visits.

During the war years, it was very difficult to get by economically. Together with Martha Meyer, who lived at the other end of the street in Rudge Hill House, Lilly Kolisko made various handicrafts: thin, finely woven scarves and necklaces from semi-precious stones. She travelled back and forth to London to sell them, and so needlework became an important source of income for her. She also sold her own jewellery just to survive. During the war, many foodstuffs were rationed, and although tickets were issued, "she did not always have the money to buy the rationed amount".[405] During this time, Martha Meyer and her family looked after Lilly Kolisko, and the children helped, for example, to pick large baskets full of dandelion flowers, which were needed for experiments with biodynamic preparations.[406]

Publications from the "Kolisko Archive"

Lilly Kolisko continued to work on various publications. As it was not possible to publish the entire book "Agriculture of Tomorrow", she published part of the book in 1943 as a special issue on capillary dynamolysis, with a description of the method and its application in various fields. Topics included various fruit preservation methods; the difference between natural and synthetic formic acid; urine tests as a diagnostic aid for diseases; the fertiliser value of various animal excretions and the diagnosis of animal diseases.[407]

Fig. 43: Geni Kolisko with her son, Andrew. Private property

She also began to publish lectures by Eugen Kolisko from the 1930s.[408] - with a request for support for further publications: "[...] We are fully aware that it is strange to ask a reader - but the times in which we live are also strange. New ideas must penetrate humanity if humanity as a whole is to survive in this terrible struggle [...].[409] From 1944 onward, she also published a series of Eugen Kolisko's lectures on nutrition, zoology, and geology. [410]

Her new friends were a great help in this work: at lectures at Wynstones School, she met two sisters, Grace and Gladys Knapp, who came from a wealthy family. From then on, she regularly received a contribution from them for her work. Gladys Knapp also helped - as did E. C. Merry and Mrs M. Dodwell - for the translations into English. Gladys Knapp, who worked as a teacher at the "Artistic Therapy College at Tuffley", was also an excellent draughtswoman and produced many illustrations and diagrams for the new publications.[411] Lilly Kolisko accepted her help "because of encounters in earlier earthly lives". [412]

During this time, she again worked on the question of how substances change during the annual festivals. There was a particular starting point for this in 1943 at Easter, as the astronomers and the church calendars each gave different dates for Easter. [413]

*

It is not known how quickly news reached Lilly Kolisko from Switzerland during the war, nor is it known when she learned of Ita Wegman's unexpected death in Arlesheim on 4 March 1943 and Elisabeth Vreede's death in Ascona, at Casa Andrea Cristoforo on 31 August 1943 - Lilly Kolisko's 54th birthday. [414] Thus, the two people with whom she had worked intensively in Central Europe and with whom she had an inner bond had crossed the threshold without her being able to be physically present during the farewell and the funeral.

A letter at the end of 1944

29. December 1944
Dear Miss Knapp,
I am deeply touched by your kind letter and generous contribution towards the publishing fund of the Kolisko Archive. It makes me very happy to see that I have got so

many good friends who care to help me with the difficult task of spreading real scientific knowledge into the present chaotic world.

I hope you do not mind that I write one letter for all of you, but I imagine you sitting here near my typewriter and so it is not just an impersonal typed letter, but a personal address.

With the help of my friends, I could publish quite a lot in the last year and perhaps will be able to [do] so again in the coming one. It is as if I would paint a huge picture using tiny mosaic stones only, putting one here, the other there and leaving it to the reader to complete the started picture. But certainly, it is better to do this - than nothing at all.

Did you by any change read an article in the "Listener" entitled "Science and Life"? It is very interesting to see how slowly conscience arises in peoples mind. Sir Frederic Kenyon is reported to have said: "Science appears to be devouring its own children". Then the article goes on: "for patently what the world needs is not fresh gadgets or fresh showers of information, but better men and women - more love, more pity, more human kindness, if we may be forgiven for using these non-scientific expressions."

"Science throws no light on the beliefs by which all of us, including the scientists, have to live and act. Our former beliefs - our judgments, that is, about the ends and purposes of human life - have mostly collapsed, which is primarily why we are where we are."

Is it not marvellous that such a thing is written now a days? Mankind needs a new knowledge of Man, of Nature and of the Universe so I think we are justified to piece it together with our little mosaic stones. Once more thank you and all the best for the coming year! L. Kolisko[415]

The Kolisko archive in Edge

At the end of 1944, Lilly Kolisko had to look for a new home for the Kolisko Archive. The chemistry room at Wynstones School, which had served as an archive for the last few years, where many lectures and talks had been held and which had been used as a workroom for preparing publications, was needed for other purposes. She had long wanted to build another building next to her own cottage in Edge for the archive. The journey to Wynstones was also long, she had to travel there by bus and felt that she was losing a lot of time for her work.

But because of the war, it was very difficult to find building materials and labourers. In this desperate situation, she received unexpected help from Grace Knapp, who had worked in the office of a building firm during the war, Messrs. Downing Rudman & Bent in Chippenham. In her memoirs in 1978, Grace Knapp wrote:

> [....] I came into contact with anthroposophy, and read and studied many of Rudolf Steiner's books, also attending conferences at Wynstones School in Gloucester. The Second World War began and food and materials became very scarce indeed, and rationing commenced. During the war, the firm I worked for carried out maintenance work to most of the Aerodromes in Wiltshire and Gloucester, and also war damage in London. Early in 1945 my sister Gladys was home at Chippenham for a week-end. She mentioned that Mrs Kolisko wanted a hut to put all her books etc. in, as she had been asked to give up her room at Wynstones School. The next day, I went to the office and asked if anyone knew of a hut for sale. The foreman of the maintenance work said he had one at Hullavington, and he would not be requiring it anymore, and it would be up for sale. He gave a description of the hut, with sizes and other details. This information was passed to Mrs Kolisko. Her reply was that it was exactly the size she was looking for. Planning permission was obtained, it was taken down, transported to and re-erected at Edge. That was during June 1945.

> Then the problem arose. How could she move all her belongings from Wynstones to Edge. This was resolved by our workmen, who had erected the hut. they moved

all her books, etc. in their lorry. These men became so attached to Mrs Kolisko, they took their wives over to visit her afterwards, and they were all interested in her publications. This was how the Kolisko Archives started at Edge at the end of the Second World War [...].[416]

This new little "building" brought a lot of hope for the future to Lilly Kolisko's extremely modest living conditions. In a letter to Grace Knapp dated 27. May 1945 she wrote:

Dear Miss Knapp,
Thank you so much for all your kindness. I am only too happy to get this hut and enclose a cheque for £117, in payment. Of course I have to borrow the money at present, but let us hope that with the help of some friends everything will be settled [...].[417]

Lilly Kolisko was very involved in the practical details of "building of the hut":

Dear Miss Knapp,
Thank you very much for your detailed letter. You sound as if the stoves would go with the hut. What have I got to pay for them? I certainly take them. It is rather nice that there is one in the small office room and one in the bigger department [...] I want to join my own library which is in the cottage with Dr. Kolisko's and that means another two shelves! That would make a bit more room in the cottage for laboratory work. [...] Thank you and your friend very much for the sketch for the lean-to. It would be marvellous to have it. And again I do not only want to store the chairs in it, but lots of laboratory equipment which is stored in the cellar of Wynstones School. [...] The colour: It would be nice to have it painted dark-green [...]." [418]

[...] How was the hut lighted? Probably it had electric light and you know I can only have oil-lamps. But this will not matter. Another thing that worries me: will water be necessary in making the concrete foundation? I must pump the rainwater and that is rather slow process. We are quite backward - no light, no water, no gas - and in the 20th century! [419]

And on 27 June she wrote: "[...] One day it rained very heavily and when I watched the water pouring down the roof, I thought that a gutter might be necessary; I could then catch the rainwater into a bucket. Do you think the gutter might be made - one of the men suggested a cement gutter, which is not so expensive [...]." [420]

The hut was finished at the end of June 1945, so Lilly Kolisko was able to put everything in the archive and her garden in order before the conference at the school at the end of July. She was one of the speakers at this conference, giving three presentations.

The "housewarming" of the hut then took place on Saturday, 28. July 1945: Lilly Kolisko gave it the name "Green Palace", and it became an important place for the archive's future effectiveness.

An association was founded to support the work of the Kolisko Archive. The archive was founded to "collect all of Eugen Kolisko's shorthand notes and lecture notes, preparing them for further publication, and translating them into English". To publish a biography of Eugen Kolisko was also one of the goals of the association. The aim was also to disseminate the knowledge gained through anthroposophy in all areas of science, art, agriculture and education and to help all those who wanted to learn about anthroposophy and its applications in life." [421]

The books from Eugen and Lilly Kolisko's library, in which most of Rudolf Steiner's writings and lectures were available in German, could be studied in the archive room. As a member of the association, you could also choose whether you wanted to pay a membership fee or help with the practical work in the archive. [422]

Other important activities were now also taking place in the "Green Palace". Ilse Ketzel wrote in her diary:

> I felt very sorry when I had to leave Stuttgart and to say goodbye to Dr Kolisko. In my last talk in Germany, I spoke to Dr. Kolisko how sad I was not to have been able to join the "Class lessons" as I was not jet long enough a member of the [Anthroposophical]Society. He replied very quietly: "Well, perhaps the opportunity will arise one day." And how right he was! - It happened about 12 years later in England, in Edge, when Mrs Kolisko started to give the Class lessons again after the war in German! - In the Kolisko Archive! [423]

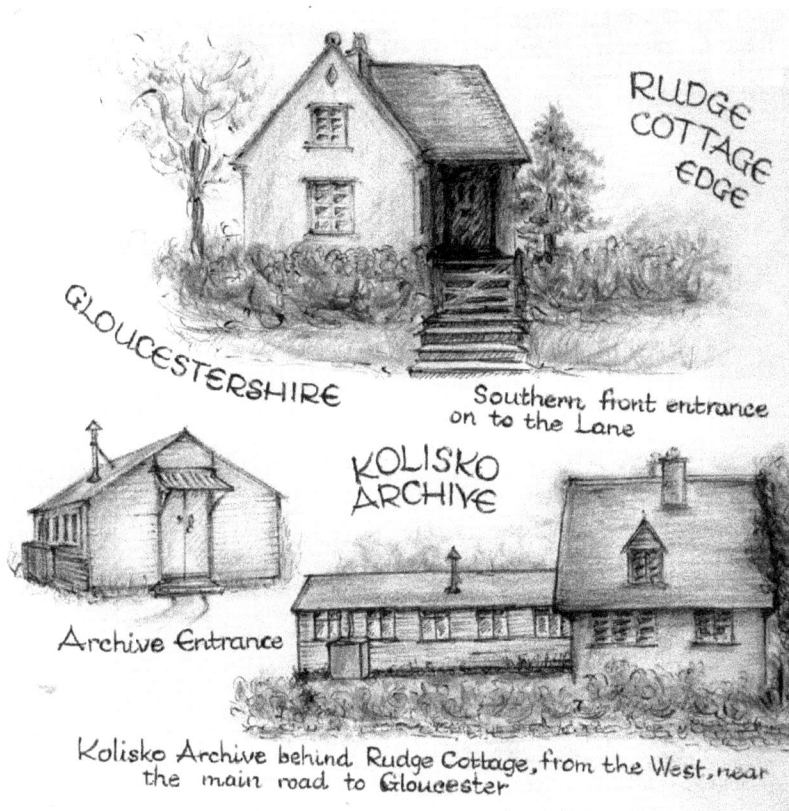

Fig. 44: Gladys Knapp: Drawing of the "Green Palace" and
Lilly Kolisko's house. From the estate of Cecil Reilly / S. T. Norway

KOLISKO ARCHIVE

RUDGE COTTAGE, EDGE
Near STROUD
Glos.

The Kolisko Archive came into being in 1940. In November, 1939, Eugen Kolisko had died suddenly, just as he and Mrs. Kolisko were about to write " Agriculture of Tomorrow," a detailed report of research work carried out over a period of 20 years. Mrs. Kolisko then appealed to a few of Dr. Kolisko's personal friends to help her financially in publishing this book. Shorthand notes of many of Dr. Kolisko's lectures on physiology, zoology, geogoly, chemistry, etc., were also available and it was hoped that some of them might be published too.

During the past 10 years the generosity of friends has enabled continuous research work to be carried on, and the following publications have been made :

1943. Advance print of Some Chapters of " Agriculture of Tomorrow." Three Fundamental Problems of the Anthroposophical Knowledge of Man.

(See Over)

I desire to become a member of the Kolisko Archive and

promise) a Donation of...
enclose) to contribute (monthly, yearly) ....

NAME ....
(Please write Title, Mr., Mrs., Miss, etc.)

ADDRESS ....

If you know any other persons who might be interested in the work of the Kolisko Archive, please give us names and addresses.

Fig. 45: A brochure from the Kolisko Archive, 1949,
from the estate of Cecil Reilly / S. T. Norway

282

25.

Research and publications after the end of the Second World War

[...] Once I had the "Biological Institute at the Goetheanum" in Stuttgart as working sphere. Here in England, I am living in a small gardener's cottage, without the benefits of electric light, or gas, or water. My experiments have to be carried out in rather primitive circumstances at present and my time is mostly dedicated to write books about the experiments carried out during many years, which I have not yet been able to publish. There would be no possibility for me to show you at present how my experiments ought to be done. But do not be discouraged. Circumstances may change. [...] [424]

Though a difficult subject, it is proposed in this publication to demonstrate that matter is not an inert mass of atoms but is capable of expressing qualities far beyond the realm into which we ordinary confine it. [425]

The publication of "Agriculture of Tomorrow"

It took another year after the end of the war for the publication to be finalised with the help of many donations from England, of Lilly - and Eugen - Kolisko's most important work.

In May 1946, in the epilogue for "Agriculture of Tomorrow", Lilly Kolisko wrote shocked by the detonation of atomic bombs at the end of the war: It was of tremendous importance that the forces of matter be liberated through potentiation and in this way bring the healing, constructive, rather than the explosive, destructive powers of the atomic bomb into the world:

> [...] It is not from oversight or from underestimation of the high technical and intellectual achievements of natural science that these sentences were written down. But where has science led us? We have a supremely perfect, pure, objective, impersonal science. Its ultimate

achievement is the release of the power of the atom: *the atomic bomb* [...].

[...] What can be done to awaken humanity? This modern science has a purely disruptive character. We are faced with the effects of pure intellect without a heart that beats for humanity. This science is like a precious jewel that has fallen into the dust.

It is up to scientists to take the first step to elevate this jewel once again. *A science of life* must be created *that places the human being at the centre, that looks at everything from the point of view of the human being and not from the point of view of science.* But man must be understood as a spiritual being and not as a combustion machine. Man, not only has a brain, an intellect, but also a soul and a spirit. Our science today is purely intellectual. Thoughts are produced and seem to run automatically. We live in illusions, in an illusory world. We are asked to accept as food something that tastes - or smells - like real food but is not. A drink that tastes of oranges (orange flavoured) or lemons (lemon flavoured) does not give us the living powers of an orange, lemon or grapefruit. We deceive ourselves with our eyes open (are they really open?). Why create false sensations? And who is carrying out all these fakes? Science, which so generously provides us with all kinds of substitutes! We cannot be offered a substitute for life. Let us therefore fill this "pure" science with a soul and turn it round from the downward path leading to destruction to the upward path leading to spiritual perfection. Force and substance are not opposite concepts, if we recognize that matter is imbued with spirit.

Let us not liberate nuclear power! Let us not unleash the evil, destructive forces on humanity - but let us strive for the good, constructive forces that are hidden in matter. Otherwise we will realise too late that we have created a boomerang that will strike back at us, destroying everything. Let us liberate the forces hidden behind the veil of substance by potentising and in this way call into action the healing, constructive and non-explosive, destructive forces. In 1923 I published the first experiments carried with the "smallest entities" by working out the information given to me by Dr Steiner in 1922. And here in England, Dr Steiner spoke at Penmaenmawr on "World Evolution and Human Evolution". In these lectures he mentioned the publication of the booklet "Physiological and Physical Proof of the Efficacy of the Smallest Entities": "In this way we have succeeded in splitting the merely material so that the truly spiritual emerges from within the merely material. For if you do not split the material into atoms, as the atomist does, but bring it into effect in its functions and forces, then you show the good will, I would say, to penetrate matter itself in order to enter the spiritual." After Dr Steiner had discussed the experiments in question in detail, he continued: "In the future, if one appreciates these research results in the right way, one will no longer seek the laws of nature merely by measuring, by weighing, i.e. merely in an atomistic way, but one will recognise how a *rhythm* already shows itself in all matter, how therefore the rhythm of the cosmos is expressed in the rhythm of natural events [...]."[426]

In June 1953, in the preface to the German edition of "Die Landwirtschaft der Zukunft", Lilly Kolisko wrote about the difficult circumstances of publication: "[...] The text of the original version was kept very short, as the shortage of paper those years-imposed restrictions. I was vividly reminded of the publication of the book 'Physiological and physical Proof of the Efficacy of the Smallest Entities' in 1923. When Dr Steiner asked me to write this book, he added with a smile: 'Write it in telegram style. At that time, funds imposed the restriction and led to this remark. Unfortunately, this restriction remained a permanent condition and stood in the way of the publication of a great deal of scientific material ..."[427]

Once again, Lilly Kolisko attached great importance to the book's matching cover. Gladys Knapp made a drawing of all the plants that were used for biodynamic preparations and all the animals from which organs were needed for their production.

*

On 13 June 1946, Lilly Kolisko wrote to Gladys Knapp again - on the new stationery: "Kolisko Archive, Rudge Cottage, Edge, Stroud, Glos.":

Dear Miss Knapp,

Thank you very much for your letter and the beautiful stamps. But you really ought not to have done it. I wanted so much to send you the book as a gift. You have helped me quite a lot in possibility to print it, when you succeeded to get me my "green palace" and so I was very glad to be able at least to send it off to you and your sister. But you seem both determined not to accept a gift without returning immediately another one. But you are right that the book is a victory book and it gives me great much joy to use your victory stamps for posting it.

It is nice that you are already awaiting my next publication. I will try to get soon the Geology lectures printed and I will start writing another big book: "The Seven Planets and the Seven Metals". That will again cost a lot of money, but I am somehow confident that it will also be done.

With warm greetings, yours sincerely, L. Kolisko

PS Mr. Shaw will get his copy as soon as more copies are coming in from the printer. [428]

Lilly Kolisko's new publication was completed in time for the "Summer Conference" in Wynstones, which enjoyed contributions by friends from Central Europe, including Madeleine van Deventer and Willem Zeylmans van Emmichoven. The title of her presentation at the conference was: "Agriculture of Tomorrow." [429]

She also sent a copy of her book to the royal family in Great Britain. Book orders were now coming in from all over the world, including South Africa. On 6 March 1947, she sent a book and a letter to V. Gorsky in South Africa (Johannesburg):

Dear Mr. V. Gorsky,

I got your kind letter of 6 February and cheque from Barclays Bank. The book has been sent to you immediately and I hope it will arrive in good condition. According to your description, you must have had quite an interesting life as mining engineer and your wife as a mineralogist working in Russia, Cyprus, Egypt and Tanganyika. I wonder in what way you wish to work in agriculture. Do you intend to take up farming?

Fig. 46: Drawing by Gladys Knapp for the cover of "Agriculture of Tomorrow".
© Biodynamic Association, Stroud

I am working here in England since 1937, but unfortunately my financial means are very weak and so the work can be carried on in a very restricted way. Still until now I have managed to carry on and even to publish a few things. The next book I would like to write is about the seven planets and the seven metals. Of course it will take some time to complete it.

With warm wishes for your own work, yours sincerely L. Kolisko[430]

In another letter to Gorsky, she wrote about the more detailed circumstances of her work:

[...]You mention you would like to see more of my own scientific work - also here I have to disappoint you. Once I had the "Biological Institute at the Goetheanum" in Stuttgart as working sphere. Here in England, I am living in a small gardener's cottage, without the benefits of electric light, or gas, or water. My experiments have to be carried out in rather primitive circumstances at present, and my time is mostly dedicated to write books about the experiments carried out during many years, which I have not yet been able to publish. There would be no possibility for me to show you at present how my experiments ought to be done. But do not be discouraged. Circumstances may change. [...][431]

At Michaelmas 1948, a detailed book review by Oskar Schmiedel on "Agriculture of Tomorrow" appeared in the Weleda-Nachrichten in Switzerland, which included the following:

This work, published some time ago in English, provides a comprehensive presentation of the results of the author L. Kolisko's decades of tireless research, insofar as they relate to and can be used in the design of a new agriculture. However, not only the research results have already been published by L. Kolisko in numerous publications are included, but also many works that have not yet appeared in print. The wealth of new ideas in this book makes it an important contribution not only to agriculture, but to cause a sensation in the widest circles. It is only to be hoped that

"LESSONS OF THE WAR"
COURSE FROM JULY 29TH — AUGUST 5TH, 1946

					8.15
MONDAY July 29th					**AMERICA** American Evolution E. E. Hollingshed Clarke, M.A.
	9.0 — 11.0	11.15 — 12.15	2.30 — 4.0	4.30 — 5.30	
TUESDAY July 30th	Eurhythmy Miss M. Bruehl or Handicraft I. Bruce-Smith, N.S.A.M. H. E. Wood, M.A. I	The Problems of To-day in Education Mr. John Baniau	Eurhythmy or Handicraft Practice I	The British Army's Experiment in Adult Education Mr. H. L. Hetherington	**BRITAIN** The responsibility of Britain in Cultural, Aesthetic and Technical Life Capt. W. O. Field, M.C., B.A.
WEDNESDAY July 31st	II	Curative Education Mr. F. Geuter	II	Exhibition of Children's Paintings Introduction by I. Bruce-Smith, N.S.A.M.	**HOLLAND** Trials of a People in War Time Zeylmans van Emmichoven, M.D.
THURSDAY August 1st	III	Psychological Effect of Shock W. Zeylmans van Emmichoven, M.D.	III	A Doctor's experiences in Czechoslovakia during the German occupation Petr Dostal, M.D.	**SWITZERLAND** The World Sanatorium M. van Deventer, M.D.
FRIDAY August 2nd	IV	Atomic Energies and Cosmic Forces George Adams, M.A.	IV	Science Discussion Group	History the War has taught us W. J. Stein, Ph.D.
SATURDAY August 3rd	IV	Agriculture of Tomorrow Mrs. L. Kolisko	4.30 — 6.0 Eurhythmy Performance in Whaddon		**GERMANY** Talk about present day position G. C. Bird, B.A. (if possible)
SUNDAY August 4th		Hopes and Fears of Humanity Mr. Adam Bittleston	3.0 — 4.0 Exhibition of Geometrical Drawings Introduction by Miss O. Whicher	4.30 — 5.30 The East looks West Impressions gained during 3½ years' captivity in the Far East Mr. George Brice	**THE WESTERN SLAVS** George Adams, M.A.
MONDAY August 5th	V	Star Events during the War Mr. W. Sucher	4.30 — 5.30 Reflections of a P.O.W. on Germany Mr. B. Mansfield	7.30 **RUSSIA** Anticipations of Russia's Future Mrs. Violet Plinche	9.0 Social Gathering Music

Fig. 47: Programme of the "Summer Conference", 29 July to 5 August 1946.
© Karl König Archive

Fig. 48: Lilly Kolisko at the Wynstones Waldorf School. © Ita Wegman Archive

it will soon appear in translation so that it can also be studied by non-English-speaking interested parties [...].[432]

Despite her extremely modest circumstances, Lilly Kolisko continued her research: capillary dynamolysis pictures followed a total solar eclipse in May 1947. Although the eclipse was not visible in Edge, she could see clear reactions in her pictures made with gold and other metal salts:

> Each eclipse produces specific effects reflected vividly in tests with filter paper through which a solution of 1 per cent, gold chloride rises, wherever the phenomenon may actually take place. Whether it happens in the arctic region or in the southern hemisphere, invariably it can be traced in our experiments without exception. What does this mean? It means that we must look upon the sun not only as a heavenly body streaming light and warmth to the earth from one particular spot, but surrounding, enveloping the whole globe with its sphere [...].[433]

Lilly Kolisko could publish these results in "Gold and the Sun" as early as the end of 1947. [434]

"Spirit in Matter"

At Easter 1948, Lilly Kolisko wrote in the foreword to another publication, "Spirit in Matter", about her intention: "Though a difficult subject, it is proposed in this publication to demonstrate that matter is not an inert mass of atoms but is capable of expressing qualities far beyond the realm into which we ordinary confine it. " [435]

The experiments carried out over many years with metal salts during various constellations and during the course of the year formed the basis of this work. Lilly Kolisko had already shown some of these pictures in her lecture at the opening of the Second Goetheanum in 1928. Over the years she expanded the work; thus it was repeatedly mentioned in her and Eugen Kolisko's letters that she did not want to travel during Christmas and the 13 Holy Nights, for example, in order to be able to continue her experiments.

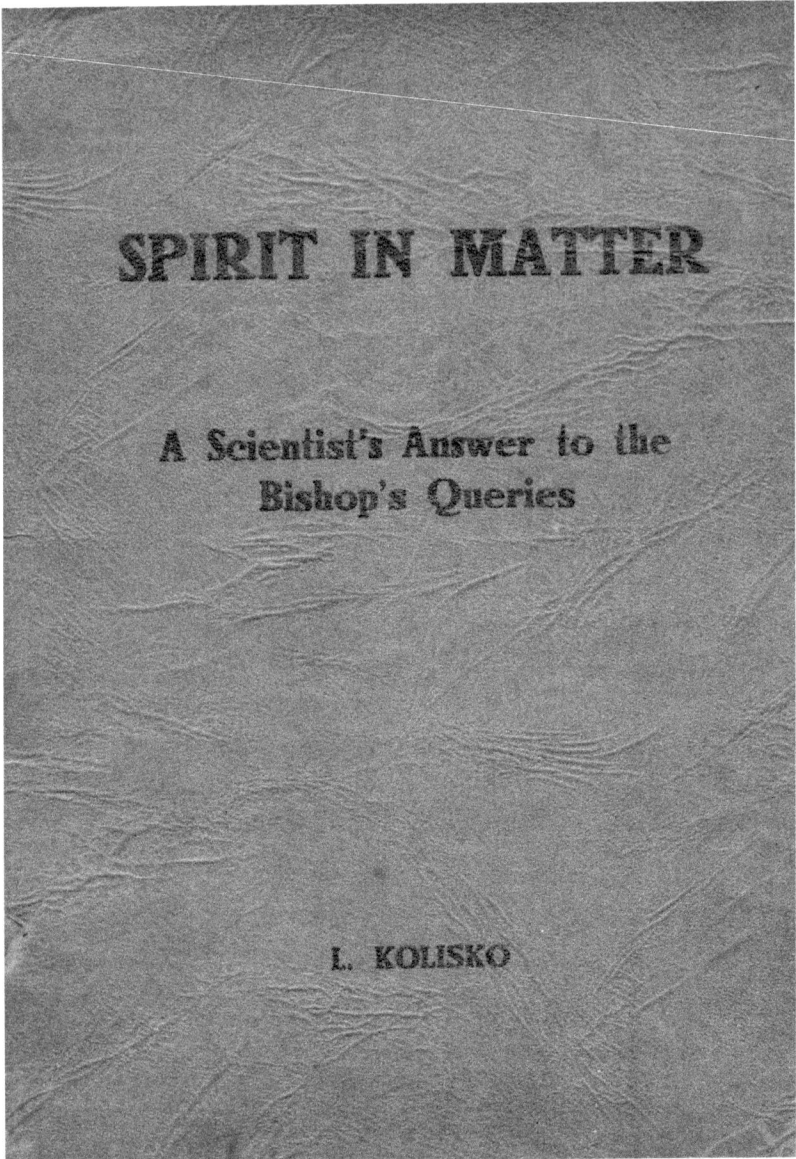

Fig. 49: Spirit in Matter, 1948. © Kolisko Archive, London

In 1943 astronomers and the Church disagreed on the date of Easter Sunday. The astronomers gave 28 March, the Church 25 April. Lilly Kolisko saw this as a good opportunity to use her capillary dynamolysis method to find out when Easter was celebrated in nature and throughout the cosmos.

In her book "Spirit in Matter" she wrote about her method in the introduction and showed pictures of earlier day and night experiments with silver nitrate and iron sulphate as well as experiments with silver nitrate over a whole year. In the fifth chapter she presented "Exceptional experiments which may be observed during the course of a year", with pictures from Easter Sunday 1927 and 1928 and from 24. June 1927, St John's Day. This was followed by the experiments at Easter 1943, at the spring equinox, on 28 March and on 25. April, which were carried out with all six planetary metals in combination with gold chloride and with gold chloride alone. After a detailed description of the various pictures, she came to the following conclusion:

> Nature celebrated Easter Festival, when we commemorate the death and resurrection of Christ, on the 28th March 1943. It was an early Easter, and everybody could see this reflected in Nature. Seeds germinated early. The trees were covered with blossoms, the Spring flowers were abundant. This glorious rising of life forces – this being in the *status nascendi* - was definitely passed a month later on the 25 April.
>
> These are scientific experiments which arouse in our souls religious feelings, and so contribute towards finding a way to unite Science, Art and Religion. A scientific test enables us to decide an argument between the Astronomer Royal and the Church authorities.[436]

She also showed two gold chloride pictures in colour reproductions from 1929 for comparison: one of Easter, made during the day, and one of Christmas (midnight). However, she had to abandon her hope of one day presenting several pictures in colour, as it was too complicated and too expensive.

Chapter 8 comprised a series of pictures made at Michaelmas 1947: Experiments with silver nitrate mixed with iron sulphate showed a surprising change from 28 to 29 September. The last picture with gold chloride and copper sulphate from Whitsun concluded her observations on the annual festivals.

The design of the cover - peach blossom colour with gold lettering - was in keeping with the last sentences of the publication:

> It seems like a miracle that earthly substances can thus reveal to us cosmic secrets. It is time to revise our conceptions of Matter based on a completely materialistic point of view. Matter is not, as we too often think, a dead, inert mass, or a whirl of atoms and electrons, which we split again and again and use for the destruction of mankind. *Matter can become a vehicle for the Spirit*, but we must see that it is a vehicle for the Spirit of Christ and not the Anti-Christ. Let us redeem Matter from its present state, where it has fallen into the abyss of materialism, and lift it up again into the realm of spirit.[437]

Publication on the biodynamic preparations

In the 1940s, a confusion arose in England between Rudolf Steiner's biodynamic compost preparations and the Q.R. method developed by Maye E. Bruce. [438]

Due to numerous inquiries, Lilly Kolisko felt it necessary to clearly describe the differences, which is why she published a special issue entitled "Agriculture of Tomorrow Preparations" at the beginning of 1949.[439]

Maye Bruce used the six plants—with the addition of honey—that Rudolf Steiner had specified for biodynamic compost preparations in her Q.R. preparation, but applied them without the corresponding sheaths and preparation processes.

In the first part of the issue, Lilly Kolisko demonstrated via the capillary dynamolysis pictures that the preparations were very different in quality: "Miss Bruce uses the plants suggested by Rudolf Steiner, but she does not carry out his instructions, so it is quite correct to say: her activator is entirely different. These two methods should not be confused." [440]

Lilly Kolisko also gave advice on the use of bio-dynamic horn manure and horn silica preparations as well as on composting and the use of compost preparations. She recommended field horsetail (Equisetum) as a prophylaxis against fungal diseases and a "seed bath"

with potentised horn manure preparation, Equisetum or oak bark preparation as a germination aid for the seeds. Her practical instructions for the greenhouse are also interesting: "It is suggested that tired soil should be moved out of the greenhouse, and arranged like a compost heap. Insert the preparations in the same way. Water the heap down regularly and apply once weekly a fine spray of diluted concentrated cow manure in the proportion of 1 liquid oz. to 1 quart of warm water. In 4 weeks the soil is refreshed and can be used again." [441]

The publication showed that Lilly Kolisko had produced the various preparations for sale: the horn manure and horn silica preparations were sold in liquid form and were to be mixed with water in the specified quantity and shaken for 5 minutes before use. [442]

She emphasised:

> It is not intended to interfere with already existing channels of distribution of the preparations made according to Rudolf Steiner's suggestions but only to help those who are not yet provided for. We wish to state, that there is no commercial interest in the background of this enterprise. The prices will be kept at a minimum, to cover expenses. Any profit, will further agricultural research work.[443]

Fig. 50: Weleda Schwäbisch Gmünd, 1953

26.

Reunions & trips to Central Europe, 1949

[...] To see the greying researcher at the lectern, working with the utmost fidelity to herself and Rudolf Steiner for decades, this experience is as moving as her results are significant. We often observe in history that revolutionary discoveries struggle with the resistance of the dull world - Frau Kolisko and her research are an example of this. Her time is yet to come - it is the future.[444]

[...] There is an indescribable cosmic security in her, but also around her - how else could this person and her work be honoured? There is no other way than to feel "guilty". Her "faithfulness" to her work, her method, is exemplary - and you recognise this when she begins to talk about her experiments in a warm, objective way.[445]

Marie Steiner died on 27 December 1948 in Beatenberg at a time when it was still impossible for Lilly Kolisko to travel to Central Europe.

It was not until the autumn of 1949, when she was already 60 years old, that Lilly Kolisko was invited to give lectures in various cities in Germany and take part in professional meetings. Jürgen von Grone, who was committed to the Anthroposophical Society and who previously been active in the anthroposophical working group with Eugen Kolisko, wrote about her visit to Stuttgart in the journal of the anthroposophical medical associations journal "Beiträge zur Erweiterung der Heilkunst" [Contributions to the Expansion of the Art of Healing]:

At the end of October, Frau Lilly Kolisko came to Stuttgart for 2 ½ weeks. Friends had been inviting her for a long time. Her visit, after some thirteen years away, was significant for our scientists, doctors and students, pharmacists, farmers and, beyond these working groups in which Frau Kolisko spoke about her work, for the entire membership.

On two "Wednesday evenings", which were also attended by friends of the Society, Lilly Kolisko gave comprehensive and at the same time very concrete insights into the experimental research she has been conducting for the past 29 years.

The large hall of the Waldorf School was packed for both slides lectures. Her friends in Stuttgart, but also the large number of those who did not know her from before, and who, because they mostly had only inadequate ideas of the far-reaching significance of her research work, were all the more eager to hear about her work, gave Frau Kolisko an open-hearted welcome. She was back for the first time from the quiet place of her work in central England to the site of her earlier work on the physically and spiritually transformed soil of central Europe.[446]

The topic chosen by Lilly Kolisko for the first evening was: "The work of the last 10 years in England". In her presentation, she focussed on her research that led to the book "The Agriculture of Tomorrow".

The second lecture was a report on her experiments with capillary dynamolysis during various Mars and Saturn constellations, in which she also presented slides from her latest book, "Spirit in Matter".

During her stay in Stuttgart, she also had several conversations with the Stuttgart Anthroposophical Medical Circle and other friends.

From 3 to 5 November 1949, Lilly Kolisko was invited to visit Weleda in Schwäbisch Gmünd and met with the scientific staff of Weleda (Wilhelm Pelikan, Oskar Schmiedel, Theodor Schwenk, Walther Cloos and partly Fritz Götte) - a detailed account of the discussions can be found in the company archives.[447] During the talks, many important questions about Rudolf Steiner's indications and their results were raised, including potentiation, animal experiments, suitable times for drug production and crystallisation experiments.

She explained her experiences and emphasised, for example, the different shaking durations for herbal and mineral substances: "For herbal substances, 2.5 minutes has proved to be a good average, for minerals 4 minutes."[448]

Very specific working methods were also discussed, such as rinsing the potentiation vessels and the order of the vessels during the experiments. She also spoke about the necessary quality of the filter paper and said that she was currently unable to obtain suitable paper in England. The account also noted:

"The favourable mood for such work is undoubtedly a cheerful, harmonious one. One should not let someone potentise who has great worries [...]. What is needed is an 'objective', 'selfless' relationship between the experimenter and his experiment, unclouded by any wishful thinking. The person's own astrality can otherwise play a disruptive role. She soon learned that some people were better suited to copper experiments, others to silver, iron, gold, or similar experiments. Not everyone is equally well-suited to all metals." The potentiser's soul mood is undoubtedly important. A prayer-like mood; a solemn and grateful mood for one's entire life. (Of course, also coupled with the necessary external alertness!) Inner interest in one's professional work, without tension, brittleness [...]."[449]

She was also asked about the origin of the capillary dynamolysis method:

It was a direct indication from Rudolf Steiner when Frau Kolisko gave him the manuscript of her work on the potency curves of the seven metals. "So you want a new task?" Dr Steiner asked Frau K. Frau K. said that she still had a lot of work to do on the previous one. Dr St.: "So, you want a new work?" Frau K.: "No, I still have enough to do." Dr Steiner: "So you want a new work?" Then he wrote down: Datura Stramonium, Atropa Belladonna, Hyoscyamus etc. Then all the German names next to it. Then he said: "Take a sheet of filter paper (drawing it as a horizontal line) and let a drop of the sap fall onto it (drawing this as a small circle above it) and observe the shapes formed." It then became apparent that the drop figures all looked very similar. "Then you might need to add some salt," said Dr. Steiner. This is how this whole method evolved. (At the same time, Dr. St. had also given her other tasks, e.g., concerning the difference between the forces of the spinal cord and those of the long bones, as well as light and dark hair, among other things.) [450]

Later Lilly Kolisko spoke about her plans and living conditions:

Frau Kolisko's next planned publications were mentioned by way of introduction: As the "Agriculture of Tomorrow", covered the most important agricultural questions, she would like to summarise her results on planets and metals in a subsequent work, and a third summarising work on potency studies is planned. Finally - or even in between - the publication of a whole series of lectures by her husband on geology, chemistry, zoology, etc., which she has available as shorthand notes. The realisation of this programme is currently extremely hampered by her difficult financial situation, which forces her to spend half of each day doing handwork (shawl weaving, etc.) to at least gradually accrue the money for a new publication. Publishing through a publisher other than her own ("Kolisko-Archive") was out of the question, as she wanted to avoid any kind of commitment. However, the establishment of a branch of the Kolisko Archive in Germany is conceivable [...] [451]

Theodor Schwenk wrote in a later recollection of this encounter:

It may have been at the end of the 1940s when Lilly Kolisko was in Schwäbisch Gmünd for talks with the scientific staff of Weleda. Since in those years I had also made many series of rising pictures of planetary constellations using the method developed by Lilly Kolisko, I now had the opportunity to show one of the most recent large series to Frau Kolisko.

The rising pictures were laid out on a long row of tables covering the entire room in the chronological order in which they were made, without any further details being entered. The pictures on filter paper of a certain type had been made in a day-night experiment, and there were several planets that were in particular angles to each other in the sky at the time, but especially Saturn and Venus on the days in question.

Lilly Kolisko walked along the pictures with great attention, stopping for a long time at some of them and also commenting or explaining one thing or another. It was clear to see how Lilly Kolisko had acquired from decades of practical experience in unbiased experiments what Goethe had called the formation of a higher organ of perception for the phenomenon in question.

Effortlessly, without prior knowledge of the constellation data, Frau Kolisko was able to characterise the planets involved in our series and the course of the various constellations of that particular day in terms of emergence, climax and fade-out in such a way that the statements were in exact agreement with the astronomical facts.

Frau Kolisko not only pointed out characteristic forms in the pictures, but also the very special colourings in the constellations, for example a very typical blue in the "interplay" of Saturn and Venus, etc. This gave us experience in reading the rising pictures into the finest details; at that time the keys to understanding them were still missing for us but would subsequently prove to be reliable. If there had been a need for "proof" in favour of the rising picture method, no one could have provided more convincing evidence.

There was a small, friendly aftermath to this lesson, which still reflects something of the nature of our researcher as a personal memory: As you can imagine, there were a lot of rising pictures laid out on the six- to eight-metre-long row of tables, which made corresponding demands on the assessment, but Frau Kolisko concluded the conversation with the remark:

"You should have made *more* pictures!" Happy to be able to say that there would have been more, but there wouldn't have been enough space to display them (especially as the time available was also limited), a second and a third time came: "You should have made *more* pictures." The practitioner had spoken![452]

From Stuttgart and Schwäbisch Gmünd, Lilly Kolisko travelled on to Ticino, Ita Wegman's last place of work. She gave lectures at Casa Andrea Cristoforo, now run by Hilma Walter, which she knew very well from the circle of "young doctors". Barbara Schaeffer wrote about this in the anthology "La Motta in Erzählungen":

"Interesting guests gradually arrived at the Casa and gave lectures, for example Lilly Kolisko, who depicted the processes in the sky with her rising pictures ..."[453]

Fig. 51: Silver nitrate - iron sulphate, one hour before a Moon-Mars conjunction. From the estate of Cecil Reilly / S. T. Norway

Fig. 52: Silver nitrate - iron sulphate, during the Moon-Mars conjunction. From the estate of Cecil Reilly / S. T. Norway

Fig. 53: Silver nitrate - iron sulphate, one hour after the conjunction moon-Mars. From the estate of Cecil Reilly / S. T. Norway

Lilly Kolisko also visited Arlesheim to give lectures and talk about her research findings.

Back in England, she worked intensively with capillary dynamolysis from the end of 1949 to visualise the various Saturn conjunctions with Mars, the Moon and the Sun. The work in the Michaelmas period in 1950 produced impressive pictures with all the planetary metals.[454]

Her plan was to publish a book on Saturn and lead, and this was also the subject of her next lectures in mid-May 1951 at the Waldorf School in Stuttgart.[455] In order to give new listeners a basic grounding, in the first two lectures she described capillary dynamolysis in detail, gave a historical account of the connection between Saturn and lead and showed pictures that confirmed this connection in many experiments. She also presented her "Horizontal Capillary Dynamolysis" with circular pictures[456] and reported on her experiences during various solar eclipses. Iris Bleisch concluded her presentation on these lectures with the following note:

> To see the greying researcher at the lectern, working with the utmost fidelity to herself and Rudolf Steiner for decades, this experience is as moving as her results are significant. We often observe in history that revolutionary discoveries struggle with the resistance of the dull world - Frau Kolisko and her research are an example of this. Her time is yet to come - it is the future.[457]

From then on, Lilly Kolisko was able to take up Rudolf Steiner's commission to hold the Class lessons in Stuttgart again and read the original version of the Class lessons given by Rudolf Steiner during her visits between 1950 and 1969.

> [...] During the Lesson she grew to greatness; she then had a powerful voice and could speak quite intimately. Emil Leinhas was deeply shaken and did not hide his emotion. He said: "This is exactly how Rudolf Steiner spoke to us in these lectures at that time."[458]

*

300

Lilly Kolisko was in Stuttgart again at the beginning of January 1952. For her lecture on 6 January, she had chosen the title: "The Christmas season experienced in a scientific experiment". The evening's presentation by Christian Lahusen provides an insight into Lilly Kolisko's working methods and her deep understanding of the reality of the Christian festivals:

> The most intimate part of the experimental work which serves to prove the influence of the stars in earth substances is that which endeavours to follow in scientific experiment what takes place in the substances during the various festive periods. The first presentation of her work in this direction was given by Frau Kolisko at the opening of the Second Goetheanum in Dornach more than 20 years ago. The work has now been ongoing for 30 years, and the small excerpt that Frau Kolisko was able to give on one evening gave the impression that behind the little that became visible, a wealth of fruits of anthroposophical-scientific work was concealed.
>
> Frau Kolisko began by explaining how the period from 24 December to 6 January is an inner journey. This path has been described by Rudolf Steiner from various angles, especially in the lecture 31 December 1914 on the World New Year, following the Norwegian "Dream Song by Olav Åsteson".
> Frau Kolisko now showed how a certain inner attitude towards scientific experimentation is necessary in order to understand the things that have initially been found from the spiritual to their fullest manifestations in earthly matter. Rudolf Steiner spoke of chemists and physicists coming one day to show how matter is constructed according to the will of Christ. The more we approach the future, the more the experiment should become a prayer. Today, devotion and reverence are very rarely found in experimentation. Every experiment poses a question. Only if the question is asked with the right attitude can we get the right answer.
> The experiments shown by Frau Kolisko allow objective phenomena to speak to us. What is expressed

in matter, how much can it reveal to us? Frau Kolisko has written about this in detail in her book "Spirit in Matter", which was published in English.

In her lecture, she first gave a brief overview of her experimental set-up, as it is known from her numerous writings, some of which were published in Germany and some in England. Rudolf Steiner had given her the task of studying the formative forces in plants and minerals. She carried out these experiments with gold chloride and silver nitrate to experimentally study the effect of solar and lunar forces on the earth. The rising pictures now showed very specific characteristic changes in the course of the year during the festive seasons. It was only by consistently observing these rising pictures from day to day over a longer period of time - for decades - that this could become increasingly clear. On the other hand, additional effects such as planetary constellations could be distinguished from the typical seasonal phenomena. The precision required for these experiments was not limited to measuring heights and colour values, but also had to be practised in a way that remains unfamiliar to modern scientific experimentation; like a work of art - such as painting – simply capturing the colours and shapes does not lead to the essential point. Something of these imponderable values became clearly tangible, as far as this is possible in a short lecture with a few dozen slides.

The different substances show very different rising pictures depending on the season in which these pictures were created. In spring, for example, copper shows itself in particularly vivid pictures, whereas iron does so in autumn. Such contrasts shed a completely new light on the connection between iron and Michaelmas, and could also provide medical suggestions, e.g. for seasonal illnesses.

Frau Kolisko, after mentioning a number of correlations that are described in detail in her publications, spoke in her lecture in particular about the changes in the substance effects at the major annual festivals. If today in Western civilisation there is an increasing tendency to regard the timing of these festivals as something historical and more or less subjective, Frau Kolisko's experiments have made it clear that these festivals are not the result of human arbitrariness but of cosmic laws acting from heaven onto earth. In the month of June around the time of St. John's Day, rising pictures are transformed from uniform, inconspicuous forms into an extraordinary wealth of shapes and colours. They exhibit complete changes in the overall design and can inspire us to view this festival even more clearly in the light to which Rudolf Steiner's lectures would suggest to us.

However, the connection between cosmic forces and a festive season becomes most vivid around Christmas. The experiments she carried out over many years during this period, of which she was able to show a selection, gave the impression that at the time of the Holy Nights from 24 December to 6 January, heavenly and earthly forces interact in a particularly intimate way. Frau Kolisko showed rising pictures of gold chloride from Christmas Eve to Epiphany from each day and night. It was remarkable to see how the nocturnal pictures showed a particular beauty and abundance, especially around Christmas, and how the daytime pictures increased in creative power at the beginning of January. It is not only that the rising pictures in the days and nights of the Christmas period are generally more expressive than the corresponding attempts of the time before and after, but a certain line of development can be seen in their sequence, which sheds new light on what Rudolf Steiner described as the inner process from the feast of the birth of Jesus to the feast of the incarnation of Christ. The forms and colours of the rising pictures of gold chloride are different again, as are the salts associated with the moon and planets around Easter time. Each festival vividly depicts something of the world forces at work in it.

However, a comprehensive and factually even more satisfactory impression can only be gained by studying the available publications in detail. However, what could not be further elaborated in the lecture in the appropriate variety was compensated for by the fact that the attitude demanded by Rudolf Steiner towards the experiment was immediately apparent in the way it was presented. In the end, against the background of three decades of life's work, it was possible to express how materialism can be overcome by learning to understand matter better than modern natural science can, by penetrating from matter to spirit, which shines through substance.[459]

Geni 1952 - Saturn and lead

Geni Kolisko had moved to Forest Row with her son Andrew. She taught eurythmy and needlework at Michael Hall School and met Knut Clunies-Ross[460], the Bothmer teacher at the school. Their wedding was celebrated in March 1952.

In November 1952, Lilly Kolisko's book "Saturn und Blei" [Saturn and Lead] was published, in the foreword to which she stated her intention to publish a seven-volume work: "This book summarises about thirty years of research work and it accessible to the public. It is not possible to publish the entire scientific material, as otherwise the size of the work, which is to consist of seven volumes, would make printing impossible."[461]

In the first part of the book, she provided basic knowledge about lead: the history of lead in the various cultural epochs, the mineralogy of lead, lead as a building material, the chemistry and geochemistry of lead, lead in relation to the human organism, lead in food and in alchemy. She also wrote about Saturn: the position of Saturn in astrology, Saturn and the science of astronomy.

In the second part, "Where is the key to the unsolved scientific problems?", she developed an anthroposophical view of lead and the effects of lead as well as Saturn with detailed quotations from Rudolf Steiner's lectures.

In the third part, she showed capillary dynamolysis experiments from 1926 to 1951 during various Saturn constellations. Many of her "horizontal capillary dynamolysis" pictures were published in this publication (later called round chromatography by other researchers). Because the printing of the book had taken so long, she was able to add pictures of the triple Saturn-Jupiter opposition from the years 1951/52.

The fourth part dealt with her crystallisation experiments with lead nitrate during a conjunction of Saturn and the moon as proof that Saturn and lead belong together. The effect of lead nitrate in "high dilutions" was also demonstrated with the help of various potentiation experiments. And because the spleen in the human organism is seen as an organ belonging to the planet Saturn, she also gave an overview of her first publication on the spleen with an evaluation of recent research.

She concluded the book by summarising all her thoughts with a lecture given by Eugen Kolisko in 1932 "about the sphere of activity in Man corresponding to lead and Saturn"[462]

After completing the manuscript of "Saturn und Blei", 63-year-old Lilly Kolisko wrote the poem Saturn-Chronos, which Norbert Glas published in his book "Lichtvolles Alter"[463], dedicated to: "Frau L. Kolisko, the great and tireless researcher".

Saturn-Chronos

The stars move in the flow of time
They encounter each other ... they pass
Connect ... and stand opposite each other;
They exchange a secret wink.

Only time allows them to meet anew

But where? ... ... and when?
The times and the places,
The possibilities are different.

Whether they tempt you to good or evil ...
Who knows? ... Who can tell? ....

Time attacks the Human like an enemy.
It sucks at the bones' marrow,
Bleaches the hair,
Slows the heartbeat.
Time makes him tired
And the limbs, oh! As heavy as lead.

Into a face's tender skin
Time etches the stories of life with merciless hands.
Time makes joints creak
Like hinges on an old door.
It makes even his blood vessels hard
And wants to break them.

Oh, time, you take away youth ...
what do you give us in return?

When You were born,
I gave myself to you freely.
You need time to grow ...
To learn ... to understand ...
to think and to contemplate
to act and
change ... .

Time is needed to forget ...
To forgive ... and make amends ...
And even more time
For the growth of your Spirit.
And when it grows,
Then the body must pass away.

The body comes to life,

and the Spirit forgets

whence it came.
The Spirit wakes up again ... ...
The body must stand still.

For a while it holds its ground and waits,
While the senses are powerful

And the Soul strives.
But when the powers of the mind rush forward,
The strength of the limbs slackens
And the step is slow.

Time sends its messengers,
To think of Him
who stands at the end,
To take you back to your home country.

The road is long ... ...
time leads to eternity,
But never stop growing within yourself.

The magazine Hippocrates

In 1952 there was an intensive discussion in Germany about the rising picture method in the journal "Hippocrates"[464]. Another researcher, Dr Curry, had claimed the method for himself or described it as his own. In May, the Stuttgart doctor Eberhard Schickler, who played a leading role in the anthroposophical medical profession in Germany at the time (and also belonged to the circle of "young physicians"), wrote to Karl König in Scotland (Camphill):

> [...] The story in Hippocrates is being worked through. This Curry is quite original in his research on radiation from the cosmos, telluric and from the earth and has no need to claim for himself the precedence of the Rising Picture Method. He started his articles because his research and instruments had been pre-emted in the magazine "Weltbild". Now Leroi has sent the publisher a supplement to the information about the origins describing this as a debt of gratitude to Frau Kolisko and Kaelin referring also to the information provided by Dr Steiner to Frau Kolisko about the use of the Goppelsroeder Capillar Analysis Method. I was able to persuade G. Husemann to approve this publication by pointing out that it is not unfavourable for us if others use this method, because then there is no longer an argument about the method, and the results are what is significant. After all, you can't get Curry to abandon the method once he's had the experience. [465]

Two weeks later, on 2 June 1952, Schickler contacted König again:

> [...] Another issue of Hippocrates will be sent to you next week (Volume 52 / Issue 10). Dobler has reserved the right to compile it for himself because Alex wanted to take up the cudgels in favour of Kolisko and Kaelin. For us it is clearly unsatisfactory, but we still have the opportunity to rectify some things in subsequent authorised editions. [466]

In July 1952, another letter from Schickler stated:

> [...] Alex Leroi and I once contacted Hippocrates to bring Frau Kolisko's work to the fore in a sufficient

manner. At first the editor *Dobler* didn't really want to do it and had an article published in which he compiled a number of authors who had worked with the rising picturee method, including Frau Kolisko. But now I have learned via Leroi that Dobler himself has found that this good Dr Curry at Lake Ammer also does not mention other authors from whom he "has" his work. In any case, Dobler reassures Frau Kolisko and Kaelin that they can have space for essays in an effort to reach a good understanding [...][467]

Lilly Kolisko's article appeared in "Hippocrates" in 1953 - 33 years after the start of her research work in Stuttgart with Rudolf Steiner. She described her capillary dynamolysis method and its application in various areas in a very factual manner:

As my research was frequently mentioned in this journal, I welcomed the invitation from the editors to comment on the issues raised.

The articles by Dr Kaelin (Arlesheim) and Dr A. Leroi (Arlesheim) show that the capillary dynamolysis method (K.D.) was developed and perfected by me over a period of about 30 years. I gave my method the name *capillary dynamolysis* to make it clear from the name that it differs from the *capillary* analysis known in scientific circles (K.A.) K.A. is a purely chemical-physical analysis: Substance solutions are allowed to rise in filter paper and their rise height is measured.

The filter paper is then used to separate the coloured zones marked on the filter paper and treat them further with various chemical reagents or subject them to spectral analysis. In this way, substances can be separated from each other that can hardly be separated by conventional physical and chemical means, or only with great difficulty. The works of Friedrich *Goppelsroeder* (Basel 1901, Capillaranalyse). [...]. I mention *Gopppelsroeder's* work in particular because he is usually consulted when discussing K.A.

My method of capillary dynamolysis differs in principle from the K.A. method in that it does not pursue the purpose of chemically analysing the substances. It is based on the view that every substance retains a certain *formative power* or *pictorial power of* its own, which can

308

be made visible. (Examples of this can be found in my publication from 1943, Capillary Dynamolysis). The K.D. uses the same medium as the K.A.: filter paper and reckons with the same phenomena of capillarity. However, *forces rather than substances are* analysed, which is why we introduced the name *dynamolysis*. In the main, we distinguish between *vertical capillary dynamolysis*, in which the filter paper is formed into cylinders and immersed in the solutions to be analysed. A detailed description of my method is not necessary at this point, as it is illustrated in the various publications (1923 - 1952). In *horizontal capillary dynamolysis*, the liquid is supplied from the centre and spreads out in a circle. An entire sheet is usually required for these experiments, so that pictures are created with an expansion of about 40 cm in diameter [...].[468]

The presentation showed the wide range of applications of capillary dynamolysis with pictures of day and night experiments with silver nitrate, pictures with plant extracts, pictures of human and animal excretions for the early diagnosis of various diseases, etc. She referred to her earlier publications to further deepen the method.

Under Lilly Kolisko's article was a note from the editors: "It was only during the printing of this issue that, thanks to the kindness of the library of the Deutsches Museum in Munich, we were able to obtain the original version of F. F. Runge's book,"Der Bildungstrieb der Stoffe" [The Formative Drive of Substances.] published in 1855. We consider it necessary that Runge's experiments be mentioned in connection with Ms. Kolisko's work, and we reserve the right to return to the topic in a later publication."

In this and other disputes, Lilly Kolisko experienced the collegial support of other anthroposophical natural scientists and doctors; however, she was also aware that the development of capillary dynamolysis and all her other work was not recognised in all anthroposophical circles. For example, she wrote about the General Assembly of the Anthroposophical Society in Dornach in 1956: "[...] Then came a report on the laboratories at the Goetheanum ... The research laboratory at the Goetheanum worked in particular with the sensitive crystallisation method of Dr. E. Pfeiffer, which had been suggested by Rudolf Steiner. Mr P. E. Schiller, who heads the Physics Laboratory, is seeking to create further experimental documentation for this. *The method of rising pictures is also used for comparisons ... Who* developed the *method of capillary dynamolysis* was characteristically concealed [...]."[469]

The agriculture of the Tomorrow

In 1953, at the request of various Swiss friends, "Agriculture of Tomorrow" was published in German under the title "Die Landwirtschaft der Zukunft". It was an enormous achievement to translate this book by Lilly Kolisko with over 450 pages.[470] Another new edition was already necessary in 1957.

In a letter to Karl König, curative educator Fried Geuter described how he experienced Lilly Kolisko during a conference in Holland in the summer of 1953:

> [...] Frau Kolisko's afternoon lecture with slides deserves great recognition. It should really have been an evening lecture. Everything was crystal clear! It is unbelievable what this lonely woman achieves - who is so lonely because of her own difficult nature.
>
> Löffler, Frau Kolisko and I lived together at Juriaanses. Löffler knows how to "release" Frau Kolisko and we spent some humorous hours together. We were able to get to know a completely different person who had matured so much and was so kind in the face of severe suffering.
>
> An indescribable cosmic security lives within her, but also around her - how else could this person and her work be emphasised? There is no other way than to feel "guilty". Her "loyalty" to her work, her method, is exemplary - and you recognise this when she begins to talk warmly and objectively about her experiments. –
>
> We often had the feeling that Dr Eugen Kolisko was present during these moments of convivial conversation. It was also an interesting sight: the broad-shouldered, somewhat rheumatically stiff body of Löffler, the white-haired, plump Lilly Kolisko - with flashing, sparkling - and often very warm mildly beaming eyes! There was something festive in this togetherness [...].[471]

Life after 1953

Lilly Kolisko's life had been divided since 1949 - between regular trips to Europe with lectures and consultations in various places and further research and publishing at home in Edge.

She earned her own income from her scientific research and publishing activities, through the support of her friends and, as before, from her handicrafts. During her travels, in addition to her lecture materials and the necessary experimental materials for capillary dynamolysis, she always had necklaces made of semi-precious stones, coloured scarves made of wool and silk and other handicrafts with her for sale and as gifts. Gladys Knapp wrote about Lilly Kolisko: "She was skilful at weaving, knitting, crocheting; was a fine seamstress; did leatherwork, bookbinding, goldsmithing, glove making [...]."[472]

When she was later offered a state pension, she did not accept it but wanted to continue living off her own income.

Lilly Kolisko lived and worked in her small house in Edge, which was relatively isolated at the time. Her friendship and co-operation with Gladys Knapp was particularly important to her. Other friends who helped Lilly Kolisko, for example in collecting plants for experiments, were Daisy and John Wood. Daisy Wood was a nursery schoolteacher, John Wood was a craft teacher and looked after the farm at the Waldorf school. He was therefore able to bring urine samples from sheep for Lilly Kolisko for capillary dynamolysis, and Lilly Kolisko gave advice on the treatment of sick animals. Daisy Wood experienced that Lilly Kolisko had an awareness of planetary movements and did not even need to look at a calendar. The study of anthroposophy accompanied her throughout her life.

In addition, she kept receiving new enquiries about lectures that Eugen Kolisko had given and continued to work with her stenographed notes.

Walter Johannes Stein visited Wynstones regularly until his death in 1955. He and Lilly Kolisko were treated with great respect at the Waldorf School. They were the "greats", the ones who had worked directly with Rudolf Steiner.

Agathe and Norbert Glas, with whom Lilly Kolisko had a lot of contact, also lived in the neighbourhood of Wynstones. He was a school doctor but also ran his public medical practice a little further on at 10 Tuffley Lane. Lilly Kolisko had a study there for her research and read the Class lessons for 14 members of the Wynstones school circle and Stroud at this location.

She wrote a letter to Norbert Glas on 6 April 1958 about the weekly Wednesday evening Class lessons: "Dear Dr Glas, please ensure that the Class members are notified again, as you did last time. I'm going to Stuttgart again in June, and on my return, we can see if anything can be arranged before the summer holidays. Best regards, L. Kolisko."[473]

Lilly Kolisko's attitude towards esoteric work at the School of Spiritual Science was filled with great seriousness: She also demanded that those who took part in the lessons commit themselves to coming to all 19 consecutive lessons[474].

Lilly Kolisko's correspondence also includes several letters and postcards from her brother and his wife. So, it seems that she had maintained these family contacts. On 16 February 1958, her brother Herman Noha died in Austria at the age of 72[475] - it is not known whether she was able to attend the funeral.

27.

Later publications

[...] It can't be a complete story, but I have nevertheless witnessed and suffered through a large piece of it. That gives me the right to write about it [...].[476]

[...] These works have become of fundamental importance for the position of anthroposophical medicine [...].[477]

Extended edition of the "smallest entities" [478]

In 1958, the "Arbeitsgemeinschaft anthroposophische Ärzte"[Working Group of Anthroposophical Doctors] in Stuttgart expressed the wish to publish an expanded edition of Lilly Kolisko's "Smallest Entities" with a summary of the over 36 years of research work.

Eberhardt Schickler wrote to Karl König on 12 June 1958: "[...] Frau Kolisko is here right now. She was delighted to hear about the collection at Comburg for the publication of her book [...]."[479]

In March 1959, after intensively working through her research material, she wrote the epilogue for this publication. In the blurb on the book cover, Gisbert Husemann emphasised the topicality and importance of her research:

> The Working Group of Anthroposophical Doctors in Stuttgart considers it an honour and a pleasure to publish this complete overview of Lilly Kolisko's research work about the "smallest entities". What began as a seed 36 years ago in Stuttgart in the then Research Institute at the Goetheanum can now appear before the world in the same place in a fully developed form.
>
> Lilly Kolisko's "germinal experiment" has become a historical symbol of a new natural science and of anthroposophy itself, which Rudolf Steiner described as "a germ in the shell of natural scientific way of thinking". When Bunsen and Kirchhoff made the

fundamental attempt at spectral analysis in 1859, materialism had reached its zenith, because the celestial bodies could now also be presented in material terms. Today, after 100 years, the pendulum has swung round. Lilly Kolisko grounded the insight of the "working of the stars in earthly substances", and she thus became the representative of a secular turning point in natural science. The discovery of the efficacy of the "smallest entities" comes from below, from the shaken potency, towards these forces. These are "research results of the very first order" (R. Steiner).

This work has become of fundamental importance for the status of anthroposophical medicine. The efficacy of a potency which the physician prescribes in his daily practice does not hover in the air of a nebulous belief, but has been experimentally proven, and this experiment has emerged from anthroposophical thinking. This fact should also inspire the homeopathic physician to come to terms with the present work, which deals with Hahnemann in detail; even if he cannot adopt the starting point as authoritative for himself, he will nevertheless find his prescription confirmed from this side. The allopathic physician will have to make a greater effort, as this book will lead him to a fundamentally different view of substance and spirit than is present in today's world view. The "rhythmised substance" conveys effects from a direction that is the opposite of that from which therapy is derived from today and will be increasingly derived from in the future, namely from sub-nature. Lilly Kolisko's work, on the other hand, speaks experimentally of cosmic effectiveness in the substance and in the remedy, not only in medicine, but also in agriculture for the healing and treatment of the earth. [480]

Subsequently, Lilly Kolisko was invited to many anthroposophical and medical conferences to give presentations on her work, including an oncological conference in Arlesheim in 1960. One participant, the ophthalmologist Michael Schad, wrote to Karl König:

[...] In Arlesheim we had a room at the Hiscia Institute, in which regular seminar lessons, eurythmy, presentations of some patients, discussion of remedies, lectures on crystallisation diagnostics, the Kaelin tests, and the planetary pictures by Miss Fyfe and Frau Kolisko were discussed. Dr Leroi led the main seminar in which he presented Steiner's many aspects of the cancer problem. In the evening the London lectures were read, in which more of the esoteric background to cancer was heard [...].[481]

At "Hiscia", the research institute of the Association for Cancer Research in Arlesheim, where capillary dynamolysis was used and where Lilly Kolisko was a frequent guest, she talked to Agnes Fyfe about the mistletoe experiments there.

The total solar eclipse in 1961

Lilly Kolisko had been busy since December 1960 with preparations and experiments for a total solar eclipse on 15 February 1961. The associated trip to northern Italy (Bordighera), where totality was to take place, and the costs for the necessary materials for the experiments and the subsequent photography of the results were made possible by a donation from Christian Gärtner and the "Arbeitsgemeinschaft der anthroposophischen Ärzte". Lilly Kolisko carried out her experiments in Bordighera without interruption from 10 February to 16 February. Subsequently, there were several important constellations, and so she also investigated the "Great Conjunction" of Jupiter and Saturn during her onward journey, as well as other constellations in Switzerland. There were two different timings for the "Great Conjunction", and to be on the safe side, she carried the work twice, with tests at both times. These experiments were carried out with different metal salts and with vertical and horizontal capillary dynamolysis. Her "Preliminary Short Report" first appeared in "Beiträgen zu eine Erweiterung der Heilkunst nach geisteswissenschaftlichen Erkenntnissen" [Contributions to an Extension of the Art of Healing based on Spiritual Scientific Insights] together with a reflection by Adalbert Stifter on the solar eclipse on 8 July 1842, and a lecture by Rudolf Steiner: "The Nature of the Solar Eclipse."[482] It was also published as a special edition by the Association of Anthroposophical Physicians with a foreword by Gisbert Husemann:

In this publication, the reader will find three very different representations summarised, all of which have the solar eclipse as their theme. L. Kolisko has visualised it as a cosmic process with her art of experimentation. Adalbert Stifter describes the sensual and moral experience he had of a solar eclipse, and Rudolf Steiner reveals its essence to us as a human-cosmic process. We enter three levels of cognition. The editor would like to express his heartfelt thanks to Frau L. Kolisko for the fact that her work and writing could be published in the centenary year of Rudolf Steiner's birth.[483]

Other anthroposophically orientated scientists were also carrying out experiments at this time. Paul Eugen Schiller wrote to Theodor Schwenk in March 1961:

Dear Dr Schwenk,
[...] Your solar eclipse experiments have had a similar fate to those carried out in Arlesheim in the Hiscia Institute, the laboratory for cancer research, Arlesheim, Kirschweg 9, and also to those carried out by Suso Vetter, Keplerwarte, Goetheanum, Dornach. The former were carried out here in Arlesheim, the latter in Florence, a centre of total darkness. It turned out, as the friends say, "less than usual".
Similar experiments were carried out in Hiscia 14 days later, i.e., during the lunar eclipse that was not visible there. I was told that these showed clearer changes. Mr Vetter observed something similar during an earlier solar eclipse, i.e. a stronger influence than during the invisible eclipse.
Frau Kolisko herself was also in Florence and conducted experiments. I do not know anything about them. The address you asked for is: Frau L. Kolisko, Rudge Cottage, Edge, near Stroud / Glos, England. It will not be easy to get an answer from her. She has become increasingly withdrawn in recent years. But I strongly advise you to try. I would also advise you to contact both Herr Vetter and Dr A. Leroi, the director of Hiscia, for further details. This collaboration will certainly lead to many things [...].[484]

The publication of the complete photographic material of her experiments was still very difficult with the technology of the time and also very expensive, indeed unaffordable for Lilly Kolisko, especially as far as the colour reproductions were concerned.

She wrote to Gladys Knapp on 31 August 1961 – her 72nd birthday:

> Dear Miss Knapp,
> You never forget! I thank you very much for your good wishes and most generous gift! [...]
> Have you watched the Moon Eclipse last Saturday? I am very satisfied with the results and will try to make some colour photographs. You know that the German friends wanted very badly to have some colour prints of the Sun Eclipse. Mr. Jennings enquired various firms, but the price is ridiculously high. Imagine, they ask for 6 pictures, that is 2 again on one side, for an edition of 2,000, the price of £445---!!! That is really too much. I also wrote to Kodak about having a transparency made, enlarged, and then a colourprint, whether they could make a true colour reproduction, because the people who make the colour blocks cannot make it directly from the transparencies.
> Kodak promised to write full details, but has not yet done so. I go on trying, but still, it will not be possible. The costs are much too high.
> Thank you once more for your kindness and warm greetings,
> Yours sincerely L. Kolisko [485]

Biography of Eugen Kolisko

During her travels to Central Europe, Lilly Kolisko experienced what was going on in the Anthroposophical Society and that many important aspects of the Society's history were being concealed or presented in a false light, including where Eugen Kolisko was concerned. Based on her own shorthand notes and notes from various meetings, letters and memories, she began to compile a biography of Eugen Kolisko at Michaelmas 1957. She chose the title for the book: "Eugen Kolisko - Ein Lebensbild. Entworfen von L. Kolisko, zugleich ein Stück Geschichte der Anthroposophischen Gesellschaft". [Eugen Kolisko- A Life Picture. Designed by L. Kolisko, also a piece of the history of the Anthroposophical Society], which made it clear that she did not just want to deal with Eugen Kolisko's life.

She wrote in the foreword:

> [...] Eugen Kolisko died on 29 November at the age of 46. The history of the Society continues into the year 1960, as some events can only be explained if the last few years are also taken into account. After Rudolf Steiner's death, many changes took place in the or Society as an organism, and gradually an almost impenetrable web of lies and slander spread about it, and I must consider it my duty to contribute my part to restoring the truth. It can't be a complete story, but I have nevertheless witnessed and suffered through a large piece of it. That gives me the right to write about it. Some things will have to be said that will be unpleasant for some people; many of a veil of illusions that has spread over events will have to be torn apart if the seriously ill organism of the Society is to be helped.
>
> We must also take into account the plight of those who joined the society later, the whole younger generation, who find it difficult, one could almost say impossible, to find their way in today's chaotic social situation.
>
> It is a society that was founded out of the spiritual, to which Rudolf Steiner linked his own destiny, and in the spiritual only the *truth* counts. It may be painful to hear the truth, but only it can help us.
>
> Wherever possible, the descriptions in this book are supported by documentary evidence. In some cases, my own shorthand notes from meetings and assemblies are used, which are essential to clarify certain events.
>
> The writing of this book was begun Michaelmas 1957 and finished on Michaelmas 1960. Some supplementary material was added later.[486]

If you read the 500 pages of this book carefully, you will also find several important sections that shed light on Lilly Kolisko's own life.

The sale of the Kolisko family's last heirloom, the small house in Bad Aussee in Austria, where she had often spent relaxing days with Eugen and Geni during their time in Stuttgart and before moving to England, made it financially possible to publish Eugen Kolisko's "Lebensbild". She used half of the purchase price for the printing and gave the other half to her daughter Geni.

Gladys Knapp was once again on hand to help with this publication:

Dear Miss Knapp,

Again, I am coming with a request! The biography of Dr, Kolisko is soon to finished and the printer wants to have the title for the binding. I have selected a linen cover and the print will be in dark red. I want it very simple of course, in a fairly bold lettering.

He sent me a sample, because there has to be made a stamp. I do not like his proposal, which you find enclosed. On the blue sheet, I have tried to make an alternative sketch. Please could you make something with a nice lettering, that could be used for the stamp? It ought to be soon. [...] [487]

Three weeks later, on 10 July 1961, Lilly Kolisko wrote to Gladys Knapp again:

I want to thank you very much indeed for your kindness to write the title again [...]. It came right on time, because I had to send the last part of the first printer's proof back and so was able also to include the title page. The next step will be that I receive the text printed on single pages- as it will appear in the finished book. This means another correction and then the actual printing will start.

If I am lucky, that book should be ready by the month of September. Then the commotion will start and I will be attacked from all sides ... Well, I think I will be able to bear it [...]. [488]

At the end of 1961, Margarete Kirchner-Bockholt informed Hilma Walter in Ascona:

Dear Hilma, we're bombarded with questions here. Schiller has been with you in the meantime. I assume you know Frau Kolisko's book and have already read some of it. Froböse then asked whether we remember that Frau Dr Steiner confronted Dr Wegman in the Carpentry workshop during the 1. Class Lesson, etc., as Frau Kolisko describes. I can't remember that, but I recall very well Boos's appearance in a Class Lesson, which is also described. Do you still remember that? Then Froböse also asked whether we can find in the estate of Frau Dr Wegman's handwritten, but corrected by Dr. [Steiner], the statutes of 3 August 1924 for the association, Allgemeine

A.G. – Dornach. I want to look it up here, but I don't know anything about it.[489]

Gisbert Husemann found other things remarkable and wrote to Karl König in March 1966:

> [...] Since [I read] Eugen Kolisko's biography, I can feel renewed karmic inklings about everything and everyone. The angel of peace was there - he went away again, but not forever.[490]

Fig. 54: *Eugen Kolisko – Ein Lebensbild 1961.*

Karl König replied in his turn:

> [...] Frau Kolisko's book deeply depressed and
> tormented me. He was certainly an angel of peace, but
> what she writes about him is probably the opposite of
> what he endeavoured to achieve. I would love to talk to
> you about it sometime.[491]

Albert Steffen and Guenther Wachsmuth from the original Executive
Council of the General Anthroposophical Society were still alive when the
book was published with its bitter social history of negative campaigns,
misjudgements and exclusions, harshness and partial cruelty. Guenther
Wachsmuth died on 1 March 1963, Albert Steffen on 13 July 1963.

28.

From Lilly Kolisko's research experiences

With reference to the action of potentised medicines, I would say that there is a range of potencies in which a medicine acts only on the metabolic human being; then comes a range where one still gets a slight metabolic effect and at the same time also an action on the rhythmic human being. Then comes a stretch where the action is purely on the rhythmic organism, and then the effects already begin to be organism of nerves and senses as well. Further still, the remedy acts exclusively on the nerve-sense organism. It is therefore difficult to specify a precise numerical indication, as the substance from which one starts on the one hand, and the individual person one is treating on the other must be considered - also the age of the person. So, you can see that there is a wide field open for the doctor to decide individually, taking account of the indications given by my graphs.[492]

Lilly Kolisko has always been able to draw on her many research projects and conversations with Rudolf Steiner to solve various problems, and gain new insights. On 6 November 1961, for example, she wrote a letter in response to Hans Krüger's questions about the potentiation and effect of potentised remedies on humans:

I can answer your enquiry about high potencies very easily. After I had published the first paper on the efficacy of the smallest entities - which extended up to the 30th decimal power - I suggested to Dr Steiner that I should go even higher with my experiments, namely up to the 60th decimal power. I explained at the time that I could imagine that further minima and maxima would then emerge, based on the results obtained so far. One could probably expect a minimum at the 28th potency and then at 42. and so on.

Dr Steiner replied: "Do it first." I did so, later showed Dr Steiner the results and asked what it would be like to potentise even higher? What should one think of the "high potencies", that are common the USA particularly, that extend into the thousands? Dr Steiner replied that people would then have to have even higher bodies on which these high potencies could exert an effect. One would exert an effect on the metabolic man with potencies up to the first minimum. After the first minimum up to the second minimum on the sensory-nerve organism. Since the human being has no other systems beyond this, it would be pointless to give him even higher potencies.

As far as your reference to the "unmanageability of the effects" is concerned, there is probably a misunderstanding. When I gave Dr Steiner the manuscript of the second work, which had been carried out up to the 60th potency, I asked him whether he would allow me to indicate precisely in the curves, in each case, a number for the first minima, second minima etc., in order to make it easier for the doctors to read the curves. Dr Steiner refused. He said: You can only specify the first minimum exactly. Then the effects merge into one another."

I have thought about this saying for a long time and today I would put it like this. Consider Rudolf Steiner's explanations about the structure of the human body and the influence of the different hierarchies. You have the reference to effectiveness of the third hierarchy between the 1ST and 21ST years of life. Then comes the next hierarchy, which works between the 21ST and 42ND year of life. However, it actually intervenes around the age of 14 and therefore overlaps. The same when the next hierarchical order takes effect at the age of 35. I would say that this relates to the potency effect:

With reference to the potency effect, I would say: There is a potency range in which a remedy acts only on the metabolic man; then follows a range in which there is still a slight metabolic effect, but there is already an effect on the rhythmic man; a stretch further on one acts purely on the rhythmic organism, then effects already begin to extend to the sensory-nerve organism. Further still, the remedy acts exclusively on the nerve-sense organism. It is therefore difficult to specify a precise numerical indication, as the substance from which one starts on the one hand, and the individual person one is treating on the other, must be considered ... also the age of the person. So you can see that there is a wide field open for the doctor to decide individually, considering the general guidance that my curves give him.

I hope that I have answered your questions.

P.S. I must add something that may of value to you. When discussing the curves, which followed my question about the high potencies, Dr Steiner said: "What is shown here in the curves as a rhythm is not only a rhythm in plant growth but also reflects such a rhythm in the human being. The 1st minimum is equivalent to sexual maturity, the second to the change that occurs in the 28th year of life in humans, and so on. To this I objected that the minimum is not always at the 14TH potency. It can also appear at the 13TH or 12TH - in some cases also at the 11TH potency. The same would also apply to the second minimum etc. Dr Steiner replied: "Well, you know that puberty can occur earlier or later in some girls." All this must be alive in the doctor's consciousness.[493]

Lilly Kolisko closely followed what was going on in the world, especially in technology and science. In her capillary dynamolysis pictures she discovered major changes in metal effectiveness after the detonation of the first atomic bomb and after human activities in space; these began in 1957 with the deployment of Earth satellites, rockets and spaceships and culminated in the subsequent moon landing in July 1969.

From 19 to 25 April 1964, a conference of anthroposophical doctors was held at Comburg, a former Benedictine monastery in Schwäbisch Hall, at which Lilly Kolisko also spoke about her experiences. Gisbert Husemann wrote to Nobert Glas and Karl König about the preparations:

[...] The demand for lectures is exceptionally high for the Comburg. The topic is: Principles and Methodology of Anthroposophical Therapy. The morning will feature thematically focused colloquia, the afternoon free therapeutic discussions. Dr. Treichler will give a lecture on artistic therapy, Frau Kolisko an evening lecture on gold, and Herr Gäch a lecture on colour therapy. I would like to thank Dr Glas very much for drawing my attention to Frau Kolisko's lecture [...].[494]

Rudolf Hauschka then reported on Lilly Kolisko's presentation:

During research into the subtle changes in matter revealed by the capillary dynamic method, an alarming symptom emerged. At one of the Easter conferences of the Association of Anthroposophical Doctors at Comburg in 1964, Frau Kolisko – the grande mistress of the capillary dynamic method – spoke on "Gold and the Signs of the Times." In previous years, she had demonstrated the connections between the planets in the rising pictures, particularly the effect of solar eclipses on gold.

During a solar eclipse, the rising picture of gold, which otherwise glows in magnificent colours, darkened and revealed grey formations. Frau Kolisko now showed rising pictures of gold from recent years, taken almost hourly and at all times of day and year, and all of them, without exception, showed the eerie dark formations that previously, in the first half of the century, only appeared during solar eclipses. She concluded her lecture by asking the doctors whether, given the changes in gold, they could still justify its use as a medicine.

The excitement among the audience was considerable, and I intended to make a contribution to the discussion the next morning, as we at WALA had also noticed this phenomenon. There was no doubt: gold had changed since the middle of the century; it had become diseased. Or in other words, the connection between solar radiation and the metal through which it passes had been disrupted. Unfortunately, in the abundance of conference topics, the discussion of this epochal phenomenon was later somewhat lost. However, in a smaller circle of doctors, the question arose as to whether diseased gold could be cured. Based on WALA's experience, we were then able to demonstrate that a renewal of the gold's quality was possible through the rhythmic treatment we apply to our herbal preparations, and that even rhythmic shaking, as is common in the potentiation process, if done correctly, results in an improvement in quality. [...][495]

29.

From the age of 70 to 80

[...] What more can one wish in life, than to have good friends? [496]

In August 1959, Lilly Kolisko turned 70 years old. Even while she was still writing the biography of Eugen Kolisko, she continued to study Rudolf Steiner's work - as she had done in all the previous years. In 1959, for example, she worked intensively with the lectures that Rudolf Steiner had given 50 years earlier, in 1908/1909: "The Apocalypse of St John", "Spiritual Hierarchies and their Reflection in the Physical World", "The Gospel of St John in Relation to the three other Gospels." [497]

Likewise, following the various planetary constellations and the annual cycle - especially the days of the Christian annual festivals with the pictures of capillary dynamolysis - was a constant endeavour. On 4 January 1966 she wrote to Gladys Knapp:

> [...] I made again the experiments during Christmastime, which partly turned out very beautiful, and I will try to make colour photographs. There it is that your donation will help me. Then I have to prepare for the complicated constellations in February including a sun eclipse. Time simply flies, but it is immensely interesting. Thank you again and my best wishes for this probably very complicated year.[498]

She continued her travels to Central Europe until 1969. Lilly Kolisko repeatedly visited friends in Arlesheim and Stuttgart, where she also hold Class lessons. Margot Rössler – founder of the Study Center for Cosmic-Artistic Zodiac Work – was among those people with whom Lilly Kolisko was in contact and exchanged ideas in Stuttgart. During her stays in Ascona, she was able to recover. She had also made new friends in the Netherlands: Since 1959, she had been visiting Jan Christiaan ten Noever de Brauw family in Zeist. Norbert Glas had introduced Lilly Kolisko to Jan Christiaan ten Noever de Brauw and Asta ten Noever de Brauw-Adolph, as they shared a common interest in anthroposophy.

Lilly Kolisko visited them for many years and often stayed there for several weeks: "She then made her rising pictures at our house, and all the windowsills were usually occupied with her pictures."[499]

In 1965, she became godmother to Asta ten Noever de Brauw-Adolph's grandchild. The house was always "sociable," and Lilly Kolisko participated in the "sessions" in which Asta ten Noever de Brauw-Adolph worked as a medium.

The astronomer and Vreede student Willy Sucher, whom Lilly Kolisko had probably known since the Stuttgart years, came to visit Edge several times after his move to America for long conversations and to exchange research experiences[500]

Lilly Kolisko's great wish was to complete the seven-volume edition on the seven metals - she had already published the first part on Saturn in 1952. In December 1962, she wrote to Gladys Knapp:

> Dear Miss Knapp,
>
> Thank you for your letter and your continuous contributions towards my work. I sincerely hope, you can really afford to do so.
>
> It was very nice in Germany and Switzerland, even the weather was beautiful. The trees were in their autumnal splendour, with bright yellow, rust, golden shimmering in the setting sun. Of course, I had to talk to many people and the friends were extremely kind to me. What more can one wish in life, than to have good friends?
>
> Now I am thinking again on a new book about the gold and would very much like to talk some of the pictures over with you. Maybe you could help me again with some drawings. I have in mind to get some sketches of the experiments which would emphasize some points, which may escape the viewer, if they are not pointed out more clearly. Do you think, this might interest you? And if so, would we see each other before your Christmas holidays? I mean before you go home. Miss Ketzel will go to Ireland on about 20th of December, and then I spend more time in Tuffley, so we could meet whenever you like to suggest.
>
> Thank you again and warm regards, Yours sincerely, L. Kolisko[501]

Fig. 55: With the godchild in Zeist, 1965. © Noever de Brauw family

Unlike during her stays in Central Europe, Lilly Kolisko lived alone in her simple house in Edge, from where it was not easy to travel even to a nearby town by public transport, especially in winter:

> [...] It would be very nice to see your sketches, but maybe we wait until the weather is better. The roads are still icy and the buses are never quite certain [...] the keys of my typewriter are icy, please excuse the typing mistakes. The postman comes at quite irregular times and all letters arrive days later. I have tried to phone you, but I could not reach you this way either. Therefore, I think we better wait a few more days until things get again more orderly.[502]

Questions about her publications kept coming in:

> 5 July 1965
> Dear Miss Knapp
> Please forgive me that I have not yet answered your very nice letter of June 21st, but you do know the reason for it, is my never-ending work. But I am still worried about your generosity. God bless you.
> Recently I got a letter from America (of course also not yet answered), from the Council for Homeopathic Research and Education, Inc. New York, in which I am asked about a translation of the book "Physiologischer und physikalischer Nachweiss der Wirksamkeit der kleinster Entitäten" into English. They even offer their help in providing funds for the translation service etc. Do you think you could slip up to see me one evening, giving me your advice how to reply adequately? It would be so nice to see you again before the summer holidays start[...].[503]

Lilly Kolisko's letter to Gladys Knapp dated 1 January 1969 is, unusually, written in German - the correspondence had previously always been conducted in English:

> Dear Miss Knapp, I'm so sorry that we haven't seen each other yet and that my Christmas present still hasn't found its way to you. I hope you know that it's only because of my work. But perhaps we could at least see each other on the last day, 6 January? I would expect you in Tuffley as usual on Monday 6 January, but perhaps not

331

until 3pm, as my experiments will be dry by then. I hope you have had a nice, quiet time.

Did you know that Mrs Glas is in hospital in Gloucester? She was taken there early in the morning on 24 December to undergo immediate surgery for appendicitis.

And now, so that you don't get out of practice, I would like to ask you to translate the following poem by Michael Bauer into English as best you can. I will also try, and then we can compare what has become of it:

Gebet um Liebe

O Gott, an Liebe mach mich überreich,
dass ich dem Brunnen an dem Wege gleich`!
Dass mir das Geben so von Herzen geht,
als wie dem Brunnen, der am Wege steht!
Und dass ich jedem geb, ob bös ob gut,
wie es der Brunnen, an dem Wege tut,
auch dass ich dienstbereit bei Tag und Nacht,
so wie der Brunnen, der am Wege wacht.
Den Überfluss der Liebe gib in mich
O Gott, das bitt ich dich.

[Prayer for love: O God, make me rich in love, that I may be like the well by the wayside That giving may come from my heart, as the well that stands by the way! And that I give to everyone, whether good or bad, like the fountain on the path. I am also ready to serve day and night, Like the well that watches by the wayside. Give me the abundance of love, O God, I beg Thee!]

That's a beautiful prayer. Don't you agree? Michael Bauer knew Dr Steiner, but was very ill and died in 1929, mourned by many anthroposophist's.

If you are unable to come on 6 January, please suggest another day.

I hope you can read this long letter. Best wishes and best wishes for the new year, also for your dear sister, your L. Kolisko[504]

She wrote another letter in German on 6 December 1969 to Gladys Knapp:

Dear Miss Knapp!
I don't know if I've told you, but please bring your

chalks with you on Monday. It would be good to make the beautiful picture accurate not only in shape but also in colour. After all, it is the colours that become paler over time.

Best regards, L. Kolisko

Please also tell your dear sister that the soup tasted very good and lasted me <u>three</u> days![505]

During these years, Gladys Knapp likely created several artistic studies of medicinal plants, including all those used for biodynamic preparations, in collaboration with Lilly Kolisko.[506] The pictures are accompanied by quotations from Rudolf Steiner's medical lectures and from the book "Fundamentals of Therapy. An Extension of the Art of Healing Through Spiritual Knowledge", written by Rudolf Steiner with Ita Wegman, as well as from books by Hilma Walter and publications by Lilly Kolisko, especially from "Agriculture of the Tomorrow." It is presumed that this work originates from a task at the beginning of her research: "Dr. Steiner gave me the task of demonstrating the connection between minerals, plants, and planets."[507] Whether the material discovered was a preparation for a new publication is unknown, but it can be assumed.

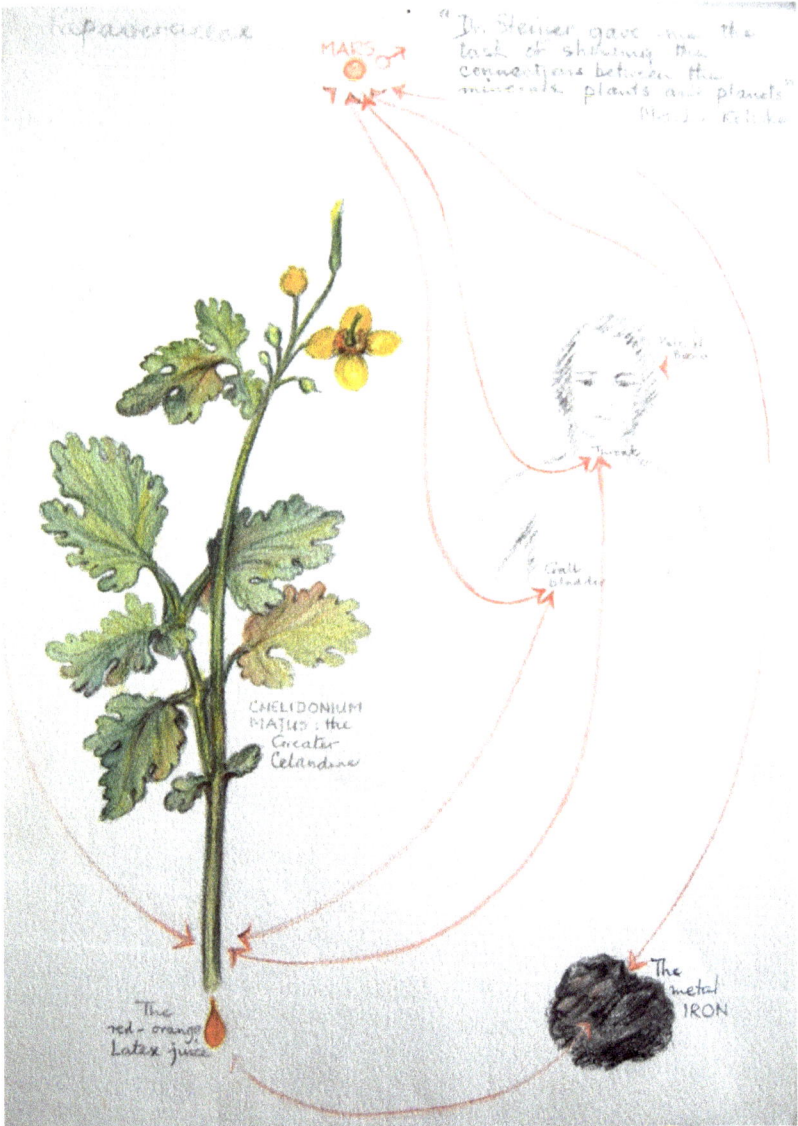

Fig. 56: Drawing by Gladys Knapp, Chelidonium Majus.
© Biodynamic Association, Stroud, England

334

30.

The last few years

"It will probably be the last time we see each other." The
countless wrinkles on her face smiled like a sun.[508]

Lilly Kolisko was no longer able to travel to Central Europe
after 1970. Her world became smaller and smaller and she
lived alone in her house. Ilse Ketzel took care of her daily
and medical needs together with Norbert Glas, and Gladys Knapp
remained a loyal friend until the end of her life.

It was time to clean up: Lilly Kolisko had used capillary dynamolysis for
diagnostic purposes during the 1960s, e.g. to examine animal faeces. She
also examined human urine and blood, and she felt an inner
responsibility towards these pictures made with blood in particular; she
knew that they could be misused if they fell into the wrong hands. In 1970
she began to destroy this research work; she burnt capillary
dynamolysis pictures and other materials.

Gisbert Husemann wrote about his last visit to Lilly Kolisko: "Miss
Gladys Knapp, her close guardian, said when I visited Frau Kolisko in
December 1970: 'Experiments in the morning, experiments in the evening,
experiments at night, two hours sleep', that was her life. When I said
goodbye to Frau Kolisko (19 December 1970) at the garden gate, she said:
'This will probably be the last time we see each other. The countless
wrinkles on her face smiled like a sun." [509]

Lilly Kolisko remained faithfully connected to the spirit of the
Goetheanum, even when her earthly work and research took place far
away from Dornach. She carried the transformation of the visible
forms of the First Goetheanum through the fire to the Foundation Stone
Meditation with her until the end of her life. Even during the last years
of her life, she worked on this meditation and attempted to translate it
into English.

Fig. 57: Lilly Kolisko with Margery Dain. © Ita Wegman Archive

Fig. 58: Lilly Kolisko. © Ita Wegman Archive

Her anthroposophical-meditative life began with the instructions from "How to Gain Knowledge of the Higher Worlds" and expanded into the meditations of the First Class of the School of Spiritual Science, in which she played a supporting role from the very beginning as a mediator of the Class lessons, from 1924 to 1969, until her journeys to Stuttgart and other places in Central Europe were no longer possible. She continued to live in this stream of meditations, and it was these that prepared the bridge across the threshold for her.

Lilly Kolisko failed to collect her milk bottles from the door in mid-November 1976, the milkman alerted the neighbours. She was then found lying on the floor - she had fallen down the stairs while fetching paraffin for the heating. She lived for another week in hospital in Gloucester - a "planetary path" - and died on Saturn's Day, Saturday 20 November 1976.

Her funeral took place at the Christian Community in Stroud and her ashes were scattered in the garden of Geni and Knut Clunies-Ross in Llandeilo, South Wales, together with Eugen Kolisko's ashes, which she had kept.

Echo

During a break in her eurythmy training in Stuttgart, Willy Stigter saw Lilly Kolisko walking up and down Kanonenweg (now Haußmannstraße) with Rudolf Steiner. When Willy Stigter came out of her lesson a few hours later, Dr Steiner and Lilly Kolisko were still walking and talking. Willy Stigter didn't see this once, but time and again. [510]

Lilly Kolisko was a delicate, almost petite figure who did not outwardly display the indomitable will that animated her. She seemed reserved, modest. She held herself very upright. Her step was determined, the pressure of her hand firm, the look in her grey eyes was expectant, scrutinising. If you spoke to her, she opened up willingly; but important matters had to be discussed.

She hated wasting time. Her answers were considered, every word weighed up. She demanded selfless dedication to the task, conscientiousness, keen observation and independence from her co-workers, but more by example than by command. Her work intensity was unusual. You could find her in her laboratory late at night or early in the morning when a series of experiments was approaching its climax.

... Doctors, astronomers, biologists, chemists, farmers and pharmacists have much to thank the deceased for. May this continue to be known and never forgotten. Her work has become a part of the history of the anthroposophical movement.[511]

It seems like a miracle that earthly substances can thus reveal to us cosmic secrets. It is time to revise our conceptions of Matter based on a completely materialistic point of view. Matter is not, as we too often think, a dead, inert mass, or a whirl of atoms and electrons, which we split again and again, and use for destruction of mankind. Matter can become a vehicle for the Spirit, but we must see that it is a vehicle for the Spirit of Christ and not the Antichrist. Let us redeem Matter from its present state, where it has fallen into the abyss of materialism, and lift it up again into the realm of Spirit.[512]

... time is needed to forget ...

To forgive ... and make amends ...

... Time sends its
messengers to think of
Him who stands at the end,
To take you back to
your home country.
The road is long ... ...

Time leads to eternity,
But never stop growing within yourself.

From the poem "Saturn and Chronos", Lilly Kolisko, 1952

Bibliography Lilly Kolisko

Publications

1922

Milzfunktion und Plättchenfrage. Der Kommende Tag AG Verlag, Stuttgart

1923

Physiologischer und physikalischer Nachweis der Wirksamkeit der kleinsten Entitäten. Der Kommende Tag AG Verlag, Stuttgart, neu herausgegeben vom Verlag am Goetheanum, Dornach 1997, mit Beiträgen von Gisbert und Friedwart Husemann

1926

Physiologischer Nachweis der Wirksamkeit kleinster Entitäten bei sieben Metallen. Wirkung von Licht und Finsternis auf das Pflanzenwachstum. Philosophisch-Anthroposophischer Verlag am Goetheanum, Dornach

1927

Sternenwirken in Erdenstoffen I. Experimentelle Studien aus dem Biologischen Institut am Goetheanum (Saturn-Sonnen-Konjunktion). Schriftenreihe der Natura, Orient-Occident-Verlag, Stuttgart
Workings of the Stars in Earthly Substances (Saturn-Sun Conjunction). Experimental Studies from the Biological Institute of the Goetheanum. Orient-Occident-Verlag, Stuttgart – The Hague – London
Sternenwirken in Erdenstoffen II. Die Sonnenfinsternis vom 29. Juni 1927. Experimentelle Studien aus dem Biologischen Institut am Goetheanum. Natura-Reihe, Orient-Occident-Verlag, Stuttgart – Den Haag – London
Workings of the Stars in Earthly Substances. The Solar Eclipse, June 29th, 1927. Experimental Studies from the Biological Institute of the Goetheanum. Orient-Occident-Verlag, Stuttgart – The Hague – London
L'Action des Astres dans les Substances Terrestres. Éditions Alice Sauerwein – Paris

1929

Sternenwirken in Erdenstoffen III. Silber und der Mond. Experimentelle Studien aus dem Biologischen Institut am Goetheanum. Schriftenreihe der Natura. Orient-Occident-Verlag, Stuttgart – Den Haag – London

1932

Sternenwirken in Erdenstoffen IV. Der Jupiter und das Zinn. Experimentelle Studien aus dem Biologischen Institut am Goetheanum. Herausgegeben von der Mathematisch-Astronomischen Sektion am Goetheanum, Dornach
Workings of the Stars in Earthly Substances IV. Jupiter and Tin. Experimental Studies from the Biological Institute of the Goetheanum, Published by the Mathematical-Astronomical Section of the Goetheanum. Dornach
Physiologischer Nachweis der Wirksamkeit kleinster Entitäten. In Fortsetzung der 1923 und 1926 hierüber erschienenen Arbeiten. Mitteilungen des Biologischen Institut am Goetheanum Nr. 1. Herausgegeben von der Medizinischen Sektion am Goetheanum, Dornach

1933

Der Mond und das Pflanzenwachstum. Mitteilungen des Biologischen Instituts am Goetheanum. Copyright Lilly Kolisko, Stuttgart, gedruckt bei Chr. Scheufele, Stuttgart

1934

Mitteilungen des Biologischen Instituts am Goetheanum No. 1 Copyright Lilly Kolisko, Stuttgart, Printed by Chr. Scheufele, Stuttgart
Mitteilungen des Biologischen Instituts am Goetheanum No. 2 Copyright Lilly Kolisko, Stuttgart, Printed by Chr. Scheufele, Stuttgart

1935

Mitteilungen des Biologischen Instituts am Goetheanum No. 3 Copyright Lilly Kolisko, Stuttgart, Printed by Chr. Scheufele, Stuttgart
Mitteilungen des Biologischen Instituts am Goetheanum No. 4 Copyright Lilly Kolisko, Stuttgart, Printed by Chr. Scheufele, Stuttgart

1936

Mitteilungen des Biologischen Instituts am Goetheanum No. 5 Gold and the

Sun. The total solar eclipse of 19 June 1936. Copyright Lilly Kolisko, Stuttgart, Printing: Chr. Scheufele, Stuttgart

Gold and the Sun. An Account of Experiments Conducted in Connection with The Total Eclipse of the Sun of 19th June 1936. Issued by the School of Spiritual Science and its Applications in Art and Life

Kristall-Gestaltungskräfte I, 12 Postkarten. Experimentelle Studien aus dem Biologischen Institut am Goetheanum

Crystalforming Process, 12 postcards in folder. Experimental studies from the Biological Institute of the Goetheanum

1943

Eugen Kolisko and Lilly Kolisko: *Capillary Dynamolysis.* Advance Print of some chapters from the second part of the book "Agriculture of Tomorrow", Kolisko Archive, Wynstones, Brookthorpe, Glos.

1946

Eugen Kolisko and Lilly Kolisko: *Agriculture of Tomorrow*, Kolisko Archive, Rudge Cottage, Edge, Stroud, England and in 1978, Kolisko Archive Publications, Bournemouth, England

Foot and Mouth Disease. Its Nature and Treatment. Reprint of Chapter XVII (Part III) from "Agriculture of Tomorrow", Kolisko Archive, Rudge Cottage, Edge, Stroud, England

1947

Gold and the Sun. An Account of Experiments Conducted in Connection with The Total Eclipse of the Sun of 20th May 1947. Kolisko Archive, Rudge Cottage, Edge, Stroud, England

1948

Spirit in Matter. A Scientist's Answer to the Bishop's Queries, 1948, Kolisko Archive, Rudge Cottage, Edge, Stroud

1949

Agriculture of Tomorrow Preparations. Agriculture of Tomorrow Publications I, Kolisko Archive, Rudge Cottage, Edge, Stroud, Glos.

1952

Sternenwirken in Erdenstoffen. Saturn und Blei. Ein Versuch, die Phänomene der Chemie, Astronomie und Physiologie zusammen zu schauen. Copyright Lilly Kolisko, Edge, Stroud, England

1953

Eugen Kolisko und Lilly Kolisko: *Die Landwirtschaft der Zukunft.* Copyright Lilly Kolisko, gedruckt bei Meyer & Cie, Schaffhausen, Schweiz, 1953 und 1957

1959

Physiologischer und physikalischer Nachweis der Wirksamkeit von kleinster Entitäten 1923–1959. Herausgegeben durch die Arbeitsgemeinschaft Anthroposophischer Ärzte, Stuttgart, Adelheidweg 4

1961

Lilly Kolisko, Adalbert Stifter, Rudolf Steiner: *Die totale Sonnenfinsternis als Experiment, Erfahrung und Wesen.* Herausgegeben von der Arbeitsgemeinschaft Anthroposophischer Ärzte, Selbstverlag, Stuttgart

The Total Eclipse of the Sun, 15 February 1961, studied in Bordighera (Northern Italy). Copyright L. Kolisko, London (See also Lilly Kolisko, Adalbert Stifter, Rudolf Steiner: *The Sun Eclipse in Experiment, as Experience, its Nature.* Bournemouth 1978, © Kolisko Archive)

Eugen Kolisko – *Ein Lebensbild.* Copyright L. Kolisko

2003

L'Agricoltura del Domina. ©Agri.Bio Edizioni, Cuneo

2017

L'Agriculture du futur. ©Éditions BioDynamie Services, France

2025

Agricultura del Mañana en cuatro tomos. ©Stay True Biodynamic Publishers, Buenos Aires

La luna e la crescita della piante. ©Agri.Bio Edizioni, Cuneo

Articles in journals

1924

"Über die Zusammenkunft der jüngeren Ärzte und Medizinstudierenden am Goetheanum, Ostern 1924." In: *Was in der Anthroposophischen Gesellschaft vorgeht.* Dornach, 1. Jahrgang, Nr. 18, 11. Mai, S. 70–72; Nr. 20 am 25. Mai, S. 78–79; Nr. 21 am 1. Juni, S. 82–83

1926/27/28

"Aus dem Biologischen Institut am Goetheanum". In: *Gäa Sophia.* Jahrbuch der naturwissenschaftlichen Sektion der Freien Hochschule für Geisteswissenschaft am Goetheanum, Dornach 1926, S. 116–117

"Vom Mysterium der Materie". In: *Natura I*, 1926/27, S. 14–21, S. 50–53, S. 177–181, S. 331–338

"Eisen und Silber ". In: *Natura I, 1926/27*, p. 79 - 85

"Zeichen der Zeit ". In: *Natura I, 1926/27*, pp. 229 - 232, pp. 385 - 387

"Erinnerungen an Rudolf Steiner ". In: *Natura I, 1926/27*, p. l76 - 282

"Etwas über die Sonnenflecken ". In: *Natura II, 1927/28*, p. 1l9 - 133

1929/30

"Der Mond und das Pflanzenwachstum." In: *Gäa Sophia.* Band IV, 1929, Landwirtschaft, Jahrbuch der naturwissenschaftlichen Sektion der Freien Hochschule am für Geisteswissenschaft am Goetheanum, S. 63–94

"Über Brot und das Quecksilber." In: *Natura III*, 1929/30, S. 25–33

1931

"Gedanken zur bevorstehenden Kalenderreform". In: *Kalender Ostern 1931 – Ostern 1932.* Mathematisch-Astronomische Sektion am Goetheanum, Dornach, S. 57–61

1953

"Aus Forschung und Wissenschaft: Kapillar-Dynamolysis. Eine spezifische Methode zu dem Studium der Gestaltungskräfte von unorganischen und organischen Substanzen. Die Anwendung dieser Methode in der Medizin, Lebensmittelkunde und Landwirtschaft." In: *Hippokrates.* Zeitschrift für praktische Heilkunde, 24. Jahrgang, Heft 5/1953, S. 130–135

1961

"Die totale Sonnenfinsternis vom 15.2.1961, beobachtet in Bordighera

(Norditalien). Vorläufiger Kurzbericht." Adalbert Stifter: "Anblick an der Welt (Sonnenfinsternis am 8.7.1842)"; Rudolf Steiner: "Das Wesen der Sonnenfinsternis" (aus: *Menschenfragen und Weltenantworten*. Dornach, 1922), *Beiträge zur Erweiterung der Heilkunst*. 1961, Nr. 5, S. 169–180

Rudolf Steiner über Potenzierung und Hochpotenzen. Aus einem Brief von Lilly Kolisko vom 6. November 1961 an Hans Krüger. In: *Merkurstab* 5/1994, S. 507

Aus einem Brief von Lilly Kolisko vom 6. November 1961 an Hans Krüger: Potenzierung und Hochpotenzen. In: *Weleda Korrespondenzblätter für Ärzte*, Nr. 138, September 1994, S. 86–87

Articles in English journals

1937

"Prelude to Scientific Research". In: *The Modern Mystic*. Vol. 1, No. 5, June 1937, p. 36 - 37

"The Influence of the Moon". In: *The Modern Mystic*. Vol. 1, No. 6, July 1937, p. 15 - 17

"Moon and Silver". In: *The Modern Mystic*. Vol. 1, No. 7, August 1937, p. 12 - 15

"Gold and the Sun,". In: *The Modern Mystic*. Vol. 1, No. 8, September 1937, p. 42- 47

"Mars and Iron, Jupiter and Tin, Saturn and Lead". In: *The Modern Mystic*. Vol. 1, No. 9, October 1937, p. 30 - 31 and 45

"Is Matter really Material? Research into the Influence of the Infinitesimal". In: *The Modern Mystic*. Vol. 1, No. 10, November 1937, p. 10 - 13

1938

"Preliminary Report of my Stay in Calcutta". In: *The Present Age*. Vol. 3, No. 5, May 1938, p. 12 - 13

1939 - 1940

"Astro-Biological Calendar: June, July, August, September, October, November, December, January 1940". In: *The Modern Mystic*. Vol 3: Astro-Biological Calendar for February, No. 1, p. 16 - 28; for March, No. 2, p. 83; for March and April, No. 3, p. 98 - 100; for May, No. 4, p. 141; for June, No. 5, p. 206; for July, No. 6, p. 235; for August, No. 7, p. 291; for September, No. 8, p. 232 and 325; for October, No. 9, p. 381; for November and December, No. 11, p. 450 - 451; for January, No.

12, p. 502 - 504 (No. 12 in January 1940)

"Moon and Silver". In: *Tomorrow. A Journal for the World Citizen of the New Age*, February 1940, p. 360 - 362

"Gold and Sun". In: *Tomorrow*. March 1940, p. 398 - 400

"Mars and Iron, Jupiter and Tin, Saturn and Lead". In: *Tomorrow*. April 1940, p. 454 - 456

About Lilly Kolisko

"'Appeal, Biological Research Institute for Investigating Cosmic Influences on Earthly Substances and Living Organisms', in Connection with the School of Spiritual Science". In: *The Modern Mystic*. London, Vol. 1, No. 9, October 1937

BENNELL, MARGARET: "Frau Kolisko: Glimpses of her Life and Work". In: *The Present Age*. London, Vol. 1, No. 12, November 1936, p. 6 - 15

HAHN HERBERT: *Von den Quellkräften der Seele*. 1974, Verlag Heide-hofbuchhandlung, Wolfgang Militz, Stuttgart, pp. 70 - 71

HAUSCHKA RUDOLF: *Wetterleuchten einer Zeitenwende*. Autobiografie, Bad Boll 1981, Vittorio Klostermann Verlag, p. 149 - 150

HUSEMANN, GISBERT: "Lili Kolisko, Werk und Wesen". In:*Beiträge zur Erweiterung der Heilkunst*. Stuttgart, 31st year, issue 2, 1978, p. 37 - 52; LILI KOLISKO: *Physiologischer und physikalischer Nachweis der Wirksamkeit kleinster Entitäten*. mit Beiträgen von Gisbert Husemann und Friedwart Husemann, 1997, Verlag am Goetheanum

HUSEMANN, GISBERT: "Lilly Kolisko – her Life and Work 1889 – 1976". In: *Archetype*. London, September 2001, ISSN 1462-8775, p. 31–48

INHETVEEN, HEIDE; SCHMITT, MATHILDE; SPIEKER, IRA: *Passion und Profession. Pionierinnen des ökologischen Landbaus*. Munich 2021, Oekom Verlag, p. 205 – 215

JURRIAANSE, THOMAS: "Lili Kolisko." In: *Mededelingen voor Leden van de Antroposofische Vereniging in Nederland*. 1977

KEYSERLINGK, ADALBERT: *Erinnerungen an frühe Forschungsarbeiten*. Dürnau 1993, Verlag der Kooperative Dürnau, S. 47–49 und S. 95–99

KNAPP. Gladys A.M.: "Spirit in Matter. The Research Work of L. Kolisko." In: *Anthroposophical Quarterly*. London, Vol, 22, No.1, Spring 1977, pp. 9–12; und in: *Journal for Anthroposophy*. London, No. 26, Autumn 1977, p. 13

KNAPP. Gladys A.M.: "In Memoriam, Frau. L. Kolisko (31st August 1889 – 10th November 1976)". In: *Star and Furrow*. Stourbridge, England, No.48, Spring 1977, p. 32

MERRY, ELEANOR C.: "A New Beginning." In: *Modern Mystics*. London, Vol. 1, No.11, December 1937, p. 30

PELIKAN, WILHELM: Zum Lebenswerk von Lilly Kolisko." In: *Das Goetheanum.* Dornach, Nr. 13/1977, S. 98–100

PENTER REINER: *Blick zurück nach vorn. Entwicklungswege und Verwandlungen des Anthroposophischen Klinisch-Therapeutischen Impulses in Unterlengenhardt.* Unterlengenhardt-Bad Liebenzell 2004, p. 19–27 und 37–38

POPPELBAUM, HERMANN: "Zur Methodik der Koliskoschen Phänomene." In: *Die Drei*. Stuttgart, VIII. Jahrgang, November 1928, p. 625–631

SCHWENK.THEODOR: "Erinnerungen zur Steigbildmethode (Kapillar-Dynamolyse). Im Gedenken an den 95. Geburtstag von Lilly Kolisko am 31.8.1984." In: *Mitteilungen aus der anthroposophischen Arbeit in Deutschland*. Stuttgart, Ausgabe 3, 1984, p. 199–200

STEIN, WALTER JOHANNES: "Frau. Kolisko's Journey to India ". In: *The Present Age.*London, Vol. 3, Nr.2, February 1938, p. 60–62

VETTER. SUSO: "Joachim Schultz (1902 – 1953), ein Pionier der astronomischen Arbeit am Goetheanum." In: *Nachrichtenblatt*. Dornach, Nr. 27, 1986, p. 111 – 113

Aus den Besprechungen mit Frau Kolisko vom 3 bis 5 November 1949 in Schwäbisch Gmünd (The Weleda employees who took part were Mr Cloos, Mr Krüger, Mr Pelikan, Dr Schmiedel, Mr Schwenk and in some cases Mr Götte). Weleda Archiv, Arlesheim

La Motta in Erzählungen 1938-2006, Brissago 2006, p. 35 and 193

Photo credits

Cover: Lilly KOLISKO with sunflowers in Stuttgart. © KOLISKO Archive, London

Fig. 1:   © Kolisko Archive, London
Fig. 2:   © Kolisko Archive, London
Fig. 3:   © Documentation at the Goetheanum, Dornach
Fig. 4:   © Kolisko Archive, London
Fig. 5:   © Rudolf Steiner Archive
Fig. 6:   From the estate of Cecil Reilly (C. R.), England / Soili Turunen (S. T.), Norway
Fig. 7:   © Kolisko Archive, London
Fig. 8:   © Kolisko Archive, London. From: Lilly Kolisko: Physiologischer und physikalischer Nachweis der Wirksamkeit kleinsten Entitäten. 1923 - 1959, p. 12
Fig. 9:   © Rudolf Steiner Archive
Fig. 10: © Verlag am Goetheanum
Fig. 11: © Rudolf Steiner Archive
Fig. 12: © Kolisko Archive, London
Fig. 13: © Hoffmann Photo Kino AG, Basel
Fig. 14: © Rudolf Steiner Archive
Fig. 15: From: DIETRICH ESTERL: *Die erste Waldorfschule Stuttgart Uhlandshöhe*. edition waldorf
Fig. 16: From: DIETRICH ESTERL: *Die erste Waldorfschule Stuttgart Uhlandshöhe*. edition waldorf
Fig. 17: © Rudolf Steiner Archive
Fig. 18: © Rudolf Steiner Archive
Fig. 19: © Rudolf Steiner Archive
Fig. 20: From the estate of Cecil Reilly (C. R.), England / Soili Turunen (S. T.), Norway
Fig. 21: From: *Modern Mystic*. February 1939
Fig. 22: © Kolisko Archive, London, from: LILLY KOLISKO: *Die Landwirtschaft der Zukunft*, p. 61
Fig. 23: From the estate of Cecil Reilly (C. R.), England / Soili Turunen (S. T.), Norway
Fig. 24: © Kolisko Archive, London

Fig. 25: © Rudolf Steiner Archive
Fig. 26: © Rudolf Steiner Archive
Fig. 27: © Rudolf Steiner Archive
Fig. 28: © Ita Wegman Archive
Fig. 29: © Ita Wegman Archive, from: *Natura.* 1926 /27, illustrations from pages 180 – 181
Fig. 30: © Kolisko Archive, London, from: *Spirit in Matter.* Offprint
Fig. 31: © Kolisko Archive, London, from: *Spirit in Matter.* Offprint
Fig. 32: © Ita Wegman Archive, from: PETER SELG: *Elisabeth Vreede*
Fig. 33: © Documentation at the Goetheanum
Fig. 34: © Kolisko Archive, London
Fig. 35: © Kolisko Archive, London
Fig. 36: From: J. E. ZEYLMANS VAN EMMICHOVEN: *Who was Ita Wegman.* Volume 2, Edition Georgenberg
Fig. 37: From: *Modern Mystic.* February 1939
Fig. 38: From the estate of Cecil Reilly (C. R.), England / Soili Turunen (S. T.), Norway
Fig. 39: From the estate of Cecil Reilly (C. R.), England / Soili Turunen (S. T.), Norway
Fig. 40: S. T. Norway
Fig. 41: Magda Meyer
Fig. 42: Private property, Edge, Stroud
Fig. 43: Magda Meyer
Fig. 44: From the estate of Cecil Reilly (C. R.), England / Soili Turunen (S. T.), Norway
Fig. 45: From the estate of Cecil Reilly (C. R.), England / Soili Turunen (S. T.), Norway
Fig. 46: © Biodynamic Association, Stroud
Fig. 47: © Karl König Archive
Fig. 48: © Ita Wegman Archive
Fig. 49: © Kolisko Archive, London
Fig. 50: © Weleda archive and scientific library of Weleda
Fig. 51: From the estate of Cecil Reilly (C. R.), England / Soili Turunen (S. T.), Norway
Fig. 52: From the estate of Cecil Reilly (C. R.), England / Soili Turunen (S. T.), Norway
Fig. 53: From the estate of Cecil Reilly (C. R.), England / Soili Turunen (S. T.), Norway
Fig. 54: © Kolisko Archive, London
Fig. 55: © Noever de Brauw family, Zeist
Fig. 56: © Biodynamic Association, Stroud, England

Fig. 57: © Ita Wegman Archive
Fig: 58: © Ita Wegman Archive
Fig: 59: © Rudolf Steiner Archive, Dornach

Notes

i   On the rising image method, see Aneta Zalecka https://kobra.uni-kassel.de/handle/113456789/1007011417189 and works by Uwe Geier, Jürgen Fritz and Beatrix Waldburger, among many others.

1   Lilly Kolisko to Marie Steiner, 14. March 1925. Rudolf Steiner Archives (RSA), Dornach.

2   GISBERT HUSEMANN: "Lilly Kolisko – Her Life and Work 1889-1976" *Archetype*, September 2001, ISSN 1462-8775 p 31.

3   EUGEN KOLISKO: Calendar (RSA)

4   GLADYS ANNIE M. KNAPP: "Spirit in Matter. The Research Work of L. Kolisko". In: *Anthroposophical Quarterly*. London, Vol. 22, No. 1, Spring 1977, p. 9 - 12.

5   HEIDE INHETVEEN, MATHILDE SCHMITT, IRA SPIEKER: *Passion und Profession. Pionierinnen des ökologischen Landbaus.* Munich 2021, p. 205.

6   GISBERT HUSEMANN: "Lilly Kolisko – Her Life and Work 1889-1976" *Archetype*, September 2001, ISSN 1462-8775 p. 31

7   LILLY KOLISKO: "Prelude to Scientific Research". In: *The Modern Mystic*. London, Vol. 1, No. 5, June 1937, p. 36.

8   Ibid.

9   LILLY KOLISKO: " Eisen und Silber". In: *Natura*. Arlesheim, vol. 1926 /27, p. 80.

10   LILLY KOLISKO: "Prelude to Scientific Research". In: *The Modern Mystic*. London, Vol. 1, No. 5, June 1937, p. 37.

11   GISBERT HUSEMANN: " Lilly Kolisko – Her Life and Work 1889-1976" *Archetype*, September 2001, ISSN 1462-8775 p. 31-32.

12   Quote from Rudolf Steiner in: GISBERT HUSEMANN: "Lilly Kolisko – Her Life and Work 1889-1976"

13   In: GLADYS ANNIE M. KNAPP: "Spirit in Matter. The Research Work of L. Kolisko". In: *Anthroposophical Quarterly*. London, Vol. 22, No. 1, Spring 1977, p. 10.

14   Ibid, p. 12.

15   GISBERT HUSEMANN: " Lilly Kolisko – Her Life and Work 1889-1976" p. 32

16   MARGARET BENNELL: "Frau Kolisko: Glimpses of her Life and Work". In: *The Present Age*. London, Vol. 1, No. 12, November 1936, p. 10.

17   Cf. HEIDE INHETVEEN, MATHILDE SCHMITT, IRA SPIEKER: *Passion und Profession, Pionierinnen des ökologischen Landbaus*, p. 207.

18   Until around 1926, she signed her letters Lilly Kolisko, later just L. Kolisko. She always signed her publications with the author's name "L. Kolisko". Only in official documents did she sign her name as "Elisabeth (Anna) Kolisko". In public she was known as "Frau (Dr) Kolisko " or in England as "Frau Kolisko". The spellings "Lili" or "Lily" come from later obituaries. It seems important to respect her own choice and to take it into account in future publications.

19   GLADYS ANNIE M. KNAPP: "Spirit in Matter. The Research Work of L. Kolisko". In: *Anthroposophical Quarterly*. London, Vol. 22, No. 1, Spring 1977, p. 10.

20   LILLY KOLISKO: "Erinnerungen an Rudolf Steiner". In: *Natura*, 1.1926 /27. p.282.

21   LILLY KOLISKO: *Eugen Kolisko - Ein Lebensbild*. Zugleich ein Stück Geschichte der Anthroposophischen Gesellschaft. Gerabronn-Grailsheim. 1961, p. 24 - 27.

22   Cf. PETER SELG: "Eugen Kolisko". In: *Anfänge anthroposophischer Heilkunst*. Dornach 2000, p. 137ff; and Eugen Kolisko: *Vom therapeutischen Charakter der Waldorfschule*. Essays and lectures. Ed. Peter Selg. Dornach 2002.

23   "Der Kommende Tag - Aktiengesellschaft zur Förderung wirtschaftlicher und geistiger Werte." [The coming Day] was an associative business enterprise in the spirit of the threefold social order, was founded on 13 March 1920 and had to be liquidated at the beginning of 1925 as a result of the general economic crisis (inflation). Cf. in detail: ALEXANDER LÜSCHER: *Der Kommende Tag AG*. Spiegel bei Bern 2005.

24   LILLY KOLISKO: *Eugen Kolisko - Ein Lebensbild*, p. 27. See also: EUGEN KOLISKO: *Das Wesen und Behandlung der Maul- und Klauenseuche*. Stuttgart 1925. EUGEN KOLISKO AND LILLY

KOLISKO: *Agriculture of Tomorrow*. Edge, England 1953, p. 371
- 399. EUGEN KOLISKO: *Das Wesen und Behandlung der Maul-
und Klauenseuche*. Edited by Peter Selg. Dornach 2001.

25 LILLY KOLISKO: "Aus dem Biologischen Institut am
Goetheanum". In: *Gäa Sophia, Yearbook of the Natural Science
Section*. Dornach 1926, p. 114 - 115.

26 Ibid, p. 115.

27 LILLY KOLISKO: "Erinnerungen an Rudolf Steiner". In: *Natura*.
1st year 1926 /27, p. 282. Reprint, anthology: Natura Verlag,
1981, Arlesheim.

28 Ibid, p. 277.

29 LILLY KOLISKO: "Aus dem Biologischen Institut am
Goetheanum" In: *Gäa Sophia, Jahrbuch der
Naturwissenschaftlichen Sektion*. 1926, p. 118 - 119.

30 Ibid. p. 116.

31 "The question arises as to what actually happens when one
prepares homeopathic remedies. It's actually in the
preparation. It's in the entire process of preparing what one is
doing. If you use silicic acid, for example, prepare it up to
high potencies; what are you actually doing? You are working
toward a certain point. In nature, everything is fundamentally
based on rhythmic processes. You are working toward a
certain zero point through a period in which the actual,
initially present effects of the substance in question emerge.
Just as, you see, when I have wealth and am constantly
spending, I reach a zero point and then go beyond the zero
point, but then receive something that is not merely not
wealth, but which goes beyond the character of wealth to debt,
it is the same when I am confronted with the substantial
properties of external substances. By remaining, so to speak,
in the effect of these substances, I finally reach the zero point,
where the effects of these substances no longer manifest
themselves in their ponderable state. If I go even further, it is
not the case that the whole story simply disappears, but rather
that the opposite emerges, and then the opposite is worked
into the surrounding medium. For me, therefore, it was always
the case that I saw the opposite effects of the substances in the
medium, in the trituration, and so on, in what is needed to

work in the homeopathic substance, the ground substance. This medium takes on a different configuration; just as I become another person when I pass from having assets to incurring debt in external social life, so substance passes into its opposite state and then imparts this opposite state, which it previously had within itself, to its environment. So if I were to say that a substance, by reducing it to ever smaller and smaller quantities, acquires certain properties, then, as I approach a certain zero point, it acquires the other property of radiating its previous properties into its surroundings and stimulating the agent with which I treat it in a corresponding manner. This stimulation can consist in directly provoking the counteraction described here, but it can also occur simply by provoking this counteraction in such a way that the substance in question is brought into a state through which it subsequently, or through which it exhibits, for example, fluorescence or phosphorescence under the influence of light. Then one has provoked the counteraction of radiating into the surroundings. These are the things that must be taken into account. It is truly not a matter of falling into the mystical, but rather of finally observing nature in its true action, observing it in such a way that we truly delve into its rhythmic course, even with reference to the properties of substances. This is, I would say, a guiding principle for truly recognizing the nature of the effects. If you potentise, you first reach a zero point. Beyond this lie opposing effects. But that is not all; rather, within the path that lies beyond this zero point, you can now again reach a zero point, which is now again a zero point for these opposing effects. Then, by going beyond this point, you can reach even higher effects, which, although again in their direction, are in the primary line, but which are of a completely different nature. Therefore, it would actually be a nice task to represent the effects that emerge during potentization in certain curves. Only one would find that these curves have to be annotated in a peculiar way. One would first have to form such a curve, and then, when one reaches the point where certain lower potencies, which are already active, cease to have an effect, and only then do higher potencies begin to have an effect, where there is a second zero point, one would have to turn at a right angle and draw the curve out into space. These are things that we will further explore in these lectures and which are intimately connected with the entire relationship between humans and all non-human nature." Rudolf Steiner: *Geisteswissenschaft und Medizin.* GA 312. Dornach 1999, p. 212-213.

32   LILLY KOLISKO: *Physiologischer Nachweis der Wirksamkeit kleinster Entitäten bei 7 Metallen.* Dornach 1926, p. 108 - 109.

33   Ibid, p. 109.

34   LILLY KOLISKO: "Aus dem Biologischen Institut am Goetheanum". In: *Gäa Sophia, Yearbook of the Natural Science Section.* 1926, p. 116 - 117.

35   ERNST LEHRS: *Gelebte Erwartung,* Stuttgart 1979, p. 193 - 194.

36   RUDOLF STEINER: *Knowledge of Higher Worlds: How is it Achieved?* GA 10. Rudolf Steiner Press 1969, p. 63 - 65.

37   ADALBERT VON KEYSERLINGK: *Developing Biodynamic Agriculture.* Reflections on early research, p. 50.

38   Among others, Friedrich Goppelsroeder and Hugo Platz.

39   LILLY KOLISKO: "Vom Mysterium der Materie". In: *Natura I,* 1926 /27, p. 20.

40   More detailed information on these experiments in: LILLY KOLISKO: *Physiologischer und physikalischer Nachweis der Wirksamkeit kleinster Entitäten.* 1923, chapter: "Versuch eines physikalischen Nachweises der Wirksamkeit kleinster Entitäten."

41   HANS KRÜGER: "Aus der Geschichte der Forschungsarbeit für die Weleda-Heilmittelherstellungen". In: Lecture: 4th Annual Conference of Weleda Pharmacists, 26th/27th March 1965 at the Goetheanum and in the Weleda Hall in Arlesheim.

42   EUGEN KOLISKO AND LILLY KOLISKO: *The Agriculture of Tomorrow.* p. 154, Edge, 1953.

43   ADALBERT GRAF VON KEYSERLINGK: *Erinnerungen an frühe Forschungsarbeiten.* 1993, p. 48.

44   ADALBERT VON KEYSERLINGK: *Developing Biodynamic Agriculture.* Reflections on early research, p. 76-77,

45   LILLY KOLISKO: *Eugen Kolisko – Ein Lebensbild,* p. 66 - 67.

46   RUDOLF STEINER: *Geisteswissenschaft und Medizin.* 15th lecture, 4 April 1920, GA 312. See also: RUDOLF STEINER: *Eine okkulte Physiologie.* 3rd lecture, Prague, 22. March 1911, GA

128.

47 LILLY KOLISKO: *Milzfunktion und Plättchenfrage.* Stuttgart 1922, p. 28.

48 LILLY KOLISKO: *Eugen Kolisko – Ein Lebensbild*, p. 28.

49 RUDOLF STEINER: *Die Grenzen der Naturerkenntnis.* GA 322.

50 RUDOLF STEINER: *Geisteswissenschaftliche Gesichtspunkte zur Therapie.* GA 313.

51 www.archivportal-d.de: University Archive Tübingen UAT 158/9923. Academic Rector's Office, student files (II).

52 MARGARET BENNELL: "Mrs L. Kolisko: Glimpses of Her Life and Work". In: *The Present Age.* Vol. 1, No. 12, Nov. 1936, p. 11.

53 LILLY KOLISKO: *Eugen Kolisko – Ein Lebensbild*, p. 67.

54 Ibid, p. 55.

55 Lilly Kolisko to Rudolf Steiner, 8 May 1922 (RSA). Wikipedia: "The autotype, also known as net etching in German, is a photographic and chemical reproduction process developed around 1880 by Georg Meisenbach in Munich for the production of printing plates for letterpress printing."

56 LILLY KOLISKO: *Eugen Kolisko – Ein Lebensbild*, p. 65.

57 V. SCHILLING (BERLIN): "Kolisko, Milzfunktion und Plättchenfrage". In: *Deutsche medizinische Wochenschrift.* Leipzig 1923, Literatur- und Verhandlungsberichte, p. 361.

58 LILLY KOLISKO: *Eugen Kolisko – Ein Lebensbild*, p. 65 - 66.

59 RUDOLF STEINER / MARIE STEINER-VON SIVERS: *Briefwechsel und Dokumente 1901 - 1925.* GA 262, Dornach 2002, p. 331. Cf. GISBERT HUSEMANN: "Friedrich Husemann zum 100. Geburtstag". In: *Beiträge zu einer Erweiterung der Heilkunst.* 40th volume, issue 4, July /August 1987, p. 208.

60 HERMANN POPPELBAUM: "L. KOLISKO, Milzfunktion und Plättchenfrage". In: *Die Drei*, No. 6, 1922, p. 481.

61 LILLY KOLISKO: *Eugen Kolisko – Ein Lebensbild*, p. 67 - 68.

62 RUDOLF STEINER: *Geistige Zusammenhänge in der Gestaltung des menschlichen Organismus.* GA 218. Dornach 1992, p. 81.

63 See PETER SELG: *Die Eröffnung des Goetheanum und die Diffamierung der Anthroposophie.* Arlesheim 2021; and PETER SELG: *"Anthroposophie als ein Streben nach Durchchristung der Welt".* Das Krisenjahr 1922 bis zum Brand des Goetheanum. Dornach 2022.

64   RUDOLF STEINER: *Das Schicksalsjahr 1923 in der Geschichte der Anthroposophischen Gesellschaft.* GA 259. Dornach 1991, p. 251.

65   Ibid, p. 70.

66   Calendar by EUGEN KOLISKO (RSA).

67   On the spleen, see RUDOLF STEINER: *Esoterische Betrachtungen karmischer Zusammenhänge.* GA 239, vol. 5, Dornach 1985, p. 15ff; RUDOLF STEINER: *Über Gesundheit und Krankheit.* GA 348, Dornach 1983, p. 206ff; RUDOLF STEINER: *Meditative Betrachtungen und Anleitungen zur Vertiefung der Heilkunst.* GA 316, Dornach 1987, p. 154 - 155.

68   LILLY KOLISKO: *Sternenwirken in Erdenstoffen, Saturn und Blei* 1952, p. 158.

69   LILLY KOLISKO: *Eugen Kolisko – Ein Lebensbild*, p. 68.

70   Ibid, p. 69.

71   RUDOLF STEINER: *Verses and meditations.* London 2004, p.143.

72   Lilly KOLISKO, Stenograms of 10. October 1928, Booklet III, p. 1 - 3, Rudolf Steiner Archive (RSA), Günther Frenz. See also: RUDOLF STEINER: *Das Verhältnis der Sternenwelt zum Menschen und des Menschen zur Sternenwelt.* Die geistige Kommunion der Menschheit GA 219. Dornach 1994, p. 177 - 195.

73   LILLY KOLISKO: *Eugen Kolisko – Ein Lebensbild*, p. 69.

74   RUDOLF STEINER: *Das Schicksalsjahr 1923 in der Geschichte der Anthroposophischen Gesellschaft.* Sitzung mit dem Dreißigerkreis, 31 January 1923. GA 259. Dornach 1991, p. 243.

75   LILLY KOLISKO: *Eugen Kolisko – Ein Lebensbild*, p. 68.

76   LILLY KOLISKO: *Physiologischer und physikalischer Nachweis der Wirksamkeit kleinster Entitäten 1923 - 1959.* Stuttgart 1959, p. 273.

77   LILLY KOLISKO: *Physiologischer und physikalischer Nachweis der Wirksamkeit kleinster Entitäten.* Stuttgart 1923. New edition Dornach 1997, with contributions by Gisbert and Friedwart Husemann.

78   RUDOLF STEINER: *Das Schicksalsjahr 1923 in der Geschichte der Anthroposophischen Gesellschaft.* GA 259. Dornach 1991, p. 243.

79   Without indication of the referee: "Physiologischer und physikalischer Nachweis der Wirksamkeit kleinster Entitäten von L. Kolisko". In: *Allgemeine Homöopathische Zeitung.* Stuttgart, December 1923, issue 3/4, p. 270.

80   DR. HENNES, MD, (Köln): "Physiologischer und physikalischer Nachweis der Wirksamkeit kleinster Entitäten, von L. Kolisko". In: *Heilkunst*. No. 4, April, p. 59.

81   Lilly Kolisko to Rudolf Steiner, 30. November 1923 (RSA).

82   LILLY KOLISKO: *Physiologischer Nachweis der Wirksamkeit kleinster Entitäten bei 7 Metallen, Wirkung von Licht und Finsternis auf das Pflanzenwachstum*. Dornach 1926.

83   RUDOLF STEINER: *The Chistmas Conference*. 1923-1924 Part II, 31. December *1923*. GA 260.

84   Ibid, 1. January 1924

85   RUDOLF STEINER: *Esoterische Unterweisungen für die Erste Klasse der Freien Hochschule für Geisteswissenschaft am Goetheanum*. GA 270. Dornach 2020, p. 372.

86   See RUDOLF STEINER: *The Christmas Conference 1923/1924*. GA 260. Cf. among others Sergej O. PROKOFIEFF / PETER SELG: *Die Weihnachtstagung und die Begründung der neuen Mysterien*. Arlesheim 2011.

87   Cf. JOHANNES KIERSCH: *Steiners individualisierte Esoterik einst und jetzt. Zur Entwicklung der Freien Hochschule für Geisteswissenschaft*. Dornach 2012, p. 116ff.

88   RUDOLF STEINER: *The Christmas Conference 1923/1924*. Part II, GA 260. 31. December 1923.

89   Ibid.

90   Ibid.

91   Ibid.

92   LILLY KOLISKO: "Aus dem biologischen Institut am Goetheanum". In: *Gäa Sophia*. 1926, p. 121 - 122.

93   Lilly and Eugen Kolisko to Ita Wegman, 16 January 1924, Ita Wegman Archive (IWA), Arlesheim.

94   RUDOLF STEINER: *Die Konstitution der Allgemeinen Anthroposophischen Gesellschaft durch die Weihnachtstagung*. GA 260a. Dornach 1987, p. 273 - 274.

95   See PETER SELG: *Rudolf Steiner und die Freie Hochschule für Geisteswissenschaft. Die Begründung der "Ersten Klasse"*. Arlesheim 2008; *Die Freie Hochschule für Geisteswissenschaft und die Michael-Schule*. Arlesheim 2014; *Die anthroposophische Weltgesellschaft und ihre Hochschule*. Dornach 2023.

96   Lilly Kolisko to Rudolf Steiner, 6. January 1924 (RSA).

97   RUDOLF STEINER: *Esoterische Unterweisungen für die Erste Klasse der Freien Hochschule für Geisteswissenschaft.* GA 270. Dornach 2020, p. 375.

98   LILLY KOLISKO: *Eugen Kolisko – Ein Lebensbild*, p. 90 – 91.

99   Ibid, p. 102 – 103.

100 Lilly Kolisko to Rudolf Steiner, 22 June 1924 (RSA).

101 Cf. among others RUDOLF STEINER: *Das Initiaten-Bewusstsein. Die wahren und falschen Wege der geistigen Forschung.* GA 243. Dornach 2004.

102 Lilly Kolisko to Ita Wegman, 17. October 1924 (IWA).

103 Lilly Kolisko to Ita Wegman, 2. November 1924 (IWA).

104 Clarita Berger to Mien Viehoff, 10. December 1924 (IWA).

105 Cf. on this group and Lilly Kolisko: PETER SELG: *Helene von Grünelius und Rudolf Steiners Kurse für junge Mediziner. Eine biographische Studie.* Dornach 2003; *Die Briefkorrespondenz der "jungen Mediziner". Eine dokumentarische Studie über die Rezeption der Rudolf Steiners "Jungmediziner"-Kursen* Dornach 2005.

106 Helene von Grunelius to Madeleine von Deventer, 5 December 1923 (IWA). Cf. a. J. E. ZEYLMANS VAN EMMICHOVEN: *Wer war Ita Wegman.* Volume 1. Heidelberg 1992, p. 142.

107 PETER SELG: *"Die Medizin muss Ernst machen mit dem geistigen Leben". Rudolf Steiners Hochschulkurse für "jungen Mediziner".* Dornach 2006.

108 Lilly Kolisko to Ita Wegman, 15 February 1924 (IWA).

109 Ibid.

110 Cf. RUDOLF STEINER: *Die Erkenntnisaufgabe der Jugend.* GA 217. Dornach 1981, p. 117 - 118.

111 Lilly Kolisko to Ita Wegman, 11 March 1924 (IWA).

112 Lilly Kolisko to Rudolf Steiner, 11 March 1924 (IWA). In: J. E. ZEYLMANS VAN EMMICHOVEN: *Wer war Ita Wegman.* Volume 1, p. 335 - 336.

113 RUDOLF STEINER: *Meditative Betrachtungen und Anleitungen zur Vertiefung der Heilkunst.* GA 316 Dornach 2022.

114 Presentation by LILLY KOLISKO: "Über die Zusammenkunft der jüngeren Ärzte und Medizinstudierenden am Goetheanum, Ostern 1924". In: *Was in der anthroposophischen Gesellschaft*

vorgeht. Dornach 1924, 1st vol., no. 21, p. 83.

115 Presentation by LILLY KOLISKO: "Über die Zusammenkunft der jüngeren Ärzte und Medizinstudierenden am Goetheanum, Ostern 1924". In: *Was in der anthroposophischen Gesellschaft vorgeht.* Dornach 1924, 1. vol., no. 18, 11. May 1924, no. 20, 25. May 1924, no. 21, 1. June 1924.

116 Presentation by LILLY KOLISKO: "Über die Zusammenkunft der jüngeren Ärzte und Medizinstudierenden am Goetheanum, Ostern 1924" from 25. May 1924. In: *Was in der anthroposophischen Gesellschaft vorgeht.* Vol. 1, No. 20, 25. May 1924, p. 78.

117 J. E. ZEYLMANS VAN EMMICHOVEN: *Wer war Ita Wegman.* Volume 1, Appendix. *"Die Mitglieder der medizinischen Sektion 1924"*, p. 347 - 348.

118 RUDOLF STEINER: *Geisteswissenschaftliche Grundlagen zum Gedeihen der Landwirtschaft.* GA 327. Dornach 1984, p. 124.

119 LILLY KOLISKO: "The Influence of the Moon". In: *The Modern Mystic.* Vol. 1, No. 6, July 1937, p. 15.

120 RUDOLF STEINER to Ita Wegman, 10 June 1924. In: J. E. ZEYLMANS VAN EMMICHOVEN: *Wer war Ita Wegman* Volume 1, p. 206.

121 RUDOLF STEINER: *Geisteswissenschaftliche Grundlagen zum Gedeihen der Landwirtschaft.* GA 327. Dornach 1984, p. 124.

122 Ibid. 1929, p. VI.

123 The Gabelsberger shorthand notes by Lilly Kolisko of the Class lessons held in Breslau on 12 and 13 June 1924 are in the archives of Perseus Verlag in Arlesheim, Switzerland. They were transcribed into plain text by Elea Gradewicz, Mexico, and published under the title *Der Meditationsweg der Michaelschule: Ergänzungsband: Die Wiederholungsstunden in Prag, Bern, Breslau, London und Dornach* by Thomas Meyer at Perseus Verlag 2016; in English in: *Rudolf Steiner's esoteric legacy. The First Class of the Michael School, Recapitulation Lessons and Mantras with two previously unpublished Lessons in Breslau.* Steiner Books / Perseus Basel 2018.

124 EUGEN KOLISKO AND LILLY KOLISKO: *Agriculture of Tomorrow.* Edge 1946; in German: *Die Landwirtschaft der Zukunft.* Edge 1953.

125 RUDOLF STEINER: *Geisteswissenschaftliche Grundlagen zum Gedeihen der Landwirtschaft.* GA 327, Dornach 1984, p. 49.

126 EUGEN KOLISKO AND LILLY KOLISKO: *Agriculture of Tomorrow* p. 18 - 19.

127 Ibid, p. 19-20. ADALBERT GRAF VON KEYSERLINGK: *Developing Biodanamic Agriculture.* Reflections on early research, p. 77.

129 EUGEN KOLISKO AND LILLY KOLISKO: *Agriculture of Tomorrow*, p. 49.

130 Lilly Kolisko to Ita Wegman, 26. February 1925 (IWA).

131 Lilly Kolisko to Rudolf Steiner, 26. February 1925 (RSA). LILLY KOLISKO reported extensively on these various underground experiments in: *Die Landwirtschaft der Zukunft*, p. 23 - 71.

132 RUDOLF STEINER: *Geisteswissenschaftliche Grundlagen zum Gedeihen der Landwirtschaft.* GA 327. Dornach 1984, p. 54 - 55.

133 ADALBERT GRAF VON KEYSERLINGK: *Developing Biodynamic Agriculture.* Reflections on early research, p. 78. See also EUGEN KOLISKO UND LILLY KOLISKO: *Die Landwirtschaft der Zukunft*, p. 76 - 79.

134 Lilly Kolisko to Ita Wegman, 30 October 1924 (IWA).

135 Ibid.

136 GISBERT HUSEMANN: *Lili Kolisko, Werk und Wesen*, p. 52.

137 LILLY KOLISKO: *Eugen Kolisko – Ein Lebensbild*, p. 100.

138 Lilly Kolisko to Ita Wegman, 30 October 1924 (IWA).

139 Ibid.

140 Dan van Bemelen to Maria Röschl, 30 October 1924 (IWA).

141 See more details about the youth circle in: RUDOLF STEINER: *Aus den Inhalten der esoterischen Stunden.* Volume 3. GA 266. Dornach 1998 (the two esoteric lessons for the esoteric youth circle, with notes on its genesis).

142 RUDOLF STEINER: *Aus den Inhalten der esoterischen Stunden.* Notizen aus der Erinnerung zum 16 October 1922. GA 266. Volume 3, Dornach 1998, p. 465 - 466.

143 F.W. ZEYLMANS VAN EMMICHOVEN to Ita Wegman, 21 November 1924 (IWA).

144 PETER SELG: *Rudolf Steiner. 1861 - 1925. Lebens- und Werkgeschichte. Volume 7: Die Freie Hochschule für Geisteswissenschaft und das Lebensende (1924 - 1925).* Arlesheim 2017.

145 LILLY KOLISKO: *Eugen Kolisko – Ein Lebensbild*, p. 103.

146 ITA WEGMAN: "Das Krankenlager, die letzten Tage und Stunden Dr. Steiners". In: *Nachrichtenblatt*, Dornach, 2. Jg. No. 16, 19. April 1925.

147 GISBERT HUSEMANN: *Lili Kolisko, Werk und Wesen*, p. 52.

148 Lilly Kolisko to Marie Steiner, 14. March 1925 (RSA): "One evening in Dornach, Geni asked me for the story of the evil dragon that is deep down in the earth and wants to destroy all people. It was already dark when we walked from Arlesheim to Dornach and the stars shone in the sky. Geni came up with the idea that the little stars watch out at night so that the evil dragon can't come up. Then she said, 'Yes, but mum, if the little stars fall asleep too, what will happen then?' - Yes, that would be very bad if the little stars just fell asleep. Then Geni says: 'Oh no, that doesn't matter. Dr Steiner is still there, and he's so good that the dragon can't do anything to us.'"

149 Marie Steiner to Eugen Kolisko, 4 April 1925 (RSA). Original letter in the appendix in: LILLY KOLISKO: *Eugen Kolisko - Ein Lebensbild*, p. 104 - 105.

150 LILLY KOLISKO: *Eugen Kolisko – Ein Lebensbild*, p. 101 - 102.

151 Ibid, p. 103.

152 Lilly Kolisko to Marie Steiner, 14 March 1925 (RSA).

153 Marie Steiner to Rudolf Steiner, 23 March 1925. In: Rudolf Steiner and Marie Steiner-von Sivers: *Briefwechsel und Dokumente 1921 - 1925*. GA 262. Dornach 2002, p. 461.

154 LILLY KOLISKO: *Eugen Kolisko – Ein Lebensbild*, p. 104.

155 Marie Steiner was referring here to the already completed takeover of the "biological department" of the "research institute" of the "Kommende Tag", which had become the "Biological Institute at the Goetheanum" (based in Stuttgart).

156 Marie Steiner to Eugen Kolisko, 4 April 1925 (RSA). Cf. LILLY KOLISKO: *Eugen Kolisko - Ein Lebensbild*, p. 104 - 105.

157 Ibid, p. 105 - 106.

158 See PETER SELG: "Eugen Kolisko". In: *Anfänge anthroposophischer Heilkunst*. Dornach 2000, p. 162ff.

159 Ibid. p. 106 - 109.

160 For the context of this declaration, however, see PETER SELG: *Die Intentionen Ita Wegmans 1925 - 1943. Zur Rehabilitierung Ita*

Wegmans. Volume 2. Arlesheim 2019, p. 50f.

161 Lilly Kolisko to Marie Steiner, 25 November 1926 (RSA).

162 Lilly Kolisko to Ita Wegman, 26 April 1925 (IWA).

163 Ernst Lehrs to Ita Wegman, 20 May 1925 (IWA).

164 Lilly Kolisko to Ita Wegman, 26 April 1925 (IWA).

165 Lilly Kolisko to Ita Wegman, 10 October 1925 (IWA).

166 ITA WEGMAN: *The Mysteries. Forest Row,* 2016, p. 57 - 58. and in "Ein Stück Mysteriengeschichte im Hinblick auf eine geisteswissenschaftliche Erweiterung der Heilkunst". In: *Natura,* III, 1928 /1929, p. 160.

167 Lilly Kolisko to Ita Wegman, 23 October 1925 (IWA).

168 Cf. J. E. ZEYLMANS VAN EMMICHOVEN: *Wer war Ita Wegman. Eine Dokumentation. Volume 3;* PETER SELG: *Die Intentionen Ita Wegmans 1925 - 1943.* Arlesheim 2019.

169 PETER SELG: *Die Intentionen Ita Wegmans 1925 - 1943,* p. 206ff.

170 Ita Wegman to Lilly Kolisko, 19 December 1925 (IWA).

171 Lilly Kolisko to Ita Wegman, 7 July 1926 (IWA).

172 Lilly Kolisko to Ita Wegman, 12 October 1926 (IWA).

173 Lilly Kolisko to Ita Wegman, 20 December 1926 (IWA).

174 Copy of a draft appeal to the members. Enclosure of a letter from Lilly Kolisko to Ita Wegman, 26 April 1925 (IWA).

175 Lilly Kolisko to Hilma Walter, 22 March 1926 (IWA).

176 Eberhard Schickler to Margarete Bockholt, 9 February 1926 (IWA).

177 Copy of a draft appeal to the members. Enclosure of a letter from Lilly Kolisko to Ita Wegman, 26. April 1925 (IWA).

178 EUGEN KOLISKO AND LILLY KOLISKO: *Agriculture of Tomorrow,* p. 193.

179 Ibid, p. 305.

180 Ibid, p. 306.

181 Lilly Kolisko to Hilma Walter, 22 March 1926 (IWA).

182 RUDOLF STEINER: *Geisteswissenschaftliche Grundlagen zum Gedeihen der Landwirtschaft.* GA 327. Dornach 1929, p. 183.

183 RUDOLF STEINER: *Geisteswissenschaftliche Gesichtspunkte zur Therapie.* GA 313, Dornach 1963, p. 129.

184 See more details in: LILLY KOLISKO: *Physiologischer und physikalischer Nachweis der Wirksamkeit kleinster Entitäten, 1923 - 1959.* XV. chapter: "Ein von Rudolf Steiner gewiesener neuer

Weg zur Zubereitung mineralischer Heilmittel: die Vegetabilisierung", p. 127 - 128.

185 Eberhard Schickler to Ita Wegman, 21 October 1925 (IWA).

186 LILLY KOLISKO: *Physiologischer Nachweis der Wirksamkeit kleinster Entitäten bei 7 Metallen, Wirkung von Licht und Finsternis auf das Pflanzenwachstum.* Dornach 1926.

187 Lilly Kolisko to Ita Wegman, 5 January 1926 (IWA).

188 Eberhard Schickler to Margarete Bockholt, 9 February 1926 (IWA).

189 Eberhard Schickler to Margarete Bockholt, 7 March 1926 (IWA).

190 Lilly Kolisko to Guenther Wachsmuth, 7 June 1926. Archive at the Goetheanum (GOEA), Dornach.

191 LILLY KOLISKO: *Eugen Kolisko – Ein Lebensbild*, p. 422 - 423.

192 Lilly Kolisko to Guenther Wachsmuth, 20 November 1926 (GOEA). See also: GUENTHER WACHSMUTH: *Die Ätherischen Bildekräfte in Kosmos, Erde und Mensch. in Weg zur Erforschung des Lebendigen.* Dornach 1924.

193 Guenther Wachsmuth to Lilly Kolisko, 24 November. 1926 (GOEA).

194 Lilly Kolisko to Guenther Wachsmuth, 19 September 1927 (GOEA).

195 Lilly Kolisko to Guenther Wachsmuth, 31 October 1927 (GOEA).

196 LILLY KOLISKO: Notebook, 26 April 1926, class X b, Monday 12 - 1 (RSA).

197 ITA WEGMAN: Vorwort zu Heft 1. in: *Natura.* 1926. Cf. a. J. E. ZEYLMANS VAN EMMICHOVEN: *Wer war Ita Wegman* Volume 2, p. 89ff.

198 Cf. PETER SELG: *Die Briefkorrespondenz der "jungen Mediziner". Eine dokumentarische Studie zur Rezeption von Rudolf Steiners "Jungmediziner"-Kursen,* p. 145ff.

199 LILLY KOLISKO: "Vom Mysterium der Materie". In: *Natura.* Volume I, 1926 /27, p. 14 - 21.

200 Ibid, p. 50 - 53.

201 Ibid., p. 177 - 181.

202 Ibid., p. 332 - 338.

203 Ibid., p. 79 - 81.

204 Ibid., p. 267 - 293.

205   Ibid., p. 279.

206   SUSO VETTER: "Joachim Schultz (1902 - 1953), ein Pionier der astronomischen Arbeit am Goetheanum". In: *Das Goetheanum / Nachrichtenblatt.* 1986, p. 113.

207   LILLY KOLISKO: *Sternenwirken in Erdenstoffen I. Experimentelle Studien aus dem Biologischen Institut am Goetheanum,* p. 5.

208   LILLY KOLISKO to Ita Wegman, 8 July 1927 (IWA).

209   SUSO VETTER: "Joachim Schultz (1902 - 1953), ein Pionier der astronomischen Arbeit am Goetheanum". In: *Das Goetheanum / Newsletter.* 1986, p. 111.

210   Ibid, p. 113.

211   LILLY KOLISKO: *Sternenwirken in Erdenstoffen I. Experimentelle Studien aus dem Biologischen Institut am Goetheanum.*

212   Ibid, p. 5.

213   Lilly Kolisko to Ita Wegman, 14. June 1927 (IWA).

214   LILLY KOLISKO: *Sternenwirken in Erdenstoffen II. Die Sonnenfinsternis of vom 29. Juni 1927. Experimentelle Studien aus dem Biologischen Institut am Goetheanum.* Stuttgart 1927. See also book review: ELISABETH VREEDE: "Sternenwirken in Erdenstoffen. Die Sonnenfinsternis vom 29. June 1927". In: *Das Goetheanum".* Dornach, Volume VII, No. 5, 1927, p. 38.

215   Lilly Kolisko to Ita Wegman, 8 July 1927 (IWA).

216   Ita Wegman to Lilly Kolisko, 19 December 1927 (IWA).

217   Eugen Kolisko to Ita Wegman, 22 December 1927 (IWA).

218   HERBERT HAHN: *Von den Quellkräften der Seele.* Stuttgart 1974, p. 70 - 71.

219   OTTO MYRBACH: "Die Rätselfrage der Natur, Sternenwirken in Erden-stoffen". In: *Neues Wiener Journal.* 7. 2.1928, No. 12.287, p. 8 - 9.

220   HERMANN POPPELBAUM: "Eine Weltkonferenz für Geisteswissenschaft in London". In: *Anthroposophie. Zeitschrift für Freies Geistesleben Stuttgart.* Number 33, 1928. Reproduced in: LILLY KOLISKO: *Eugen Kolisko - Ein Lebensbild,* p. 189 - 190.

221   See announcement in *Natura* 1926 /27, p. 344 and 388.

222   Lilly Kolisko to Ita Wegman, 20 June 1927 (IWA).

223   LILLY KOLISKO: *Workings of the Stars in Earthly Substances. Experimental Studies from the Biological Institute of the Goetheanum.* Stuttgart 1927.

224 EUGEN KOLISKO: "Vom Homöopathischen Kongress in London". In: *Natura II*. 1927/28, p. 128.

225 Margarete Bockholt to Hilma Walter, 27 July 1927 (IWA).

226 OTTO MYRBACH: "Die Rätselfrage der Natur, Sternenwirken in Erdenstoffen". In: *Neues Wiener Journal*. 7. February 1928, No. 12.287, p. 8 - 9.

227 Ibid, p. 9.

228 On the "Worldconference on Spiritual Science and its Practical Applications for the Well-Being of Humanity" see PETER SELG: *Die Intentionen Ita Wegmans 1925 - 1943*, p. 90ff.

229 HERMANN POPPELBAUM: "Eine Weltkonferenz für Geisteswissenschaft in London". In: *Anthroposophie*, No. 33, 1928. Reproduced in: LILLY KOLISKO: *Eugen Kolisko - Ein Lebensbild*, p. 189 - 190.

230 Lilly Kolisko to Guenther Wachsmuth, 9 August 1928 (GOEA).

231 Lilly Kolisko to Ernst Schmid-Curtius, 8 August 1928 (GOEA).

232 Lilly Kolisko to Ita Wegman and Guenther Wachsmuth, 13 August 1928 (IWA and GOEA).

233 Margarete Bockholt to Eugen Kolisko, 5 September 1928 (IWA).

234 Margarete Bockholt to Eugen Kolisko, 5 September 1928 (IWA).

235 Ita Wegman to Lilly Kolisko, 8 September 1928 (IWA).

236 Lilly Kolisko to Ita Wegman, 11 September 1928 (IWA).

237 Lilly Kolisko to Ita Wegman, 21 September 1928 (IWA).

238 LILLY KOLISKO: *Eugen Kolisko - Ein Lebensbild*, p. 191.

239 Lilly Kolisko to Guenther Wachsmuth, 8 September 1928 (GOEA).

240 LILLY KOLISKO: *Eugen Kolisko - Ein Lebensbild*, p. 190 - 192.

241 HERMANN POPPELBAUM: "Zur Methodik der Kolisko'schen Phanomene". In: *Die Drei*. VIII. Jahrgang, November 1928, p. 625 - 631.

242 Ibid, pp. 630 - 631.

243 On these disputes at the end of the 1920s see LILLY KOLISKO: *Eugen Kolisko - Ein Lebensbild,* p. 179ff.; and PETER SELG: *Die Intentionen Ita Wegmans 1925 - 1943,* p. 87ff.

244 Ita Wegman to Lilly Kolisko, 28 October 1928 (IWA).

245 Lilly Kolisko to Ita Wegman, 6 November 1928 (IWA).
246 Lilly Kolisko to Ita Wegman, 24 November 1928 (IWA).
247 Ita Wegman to Lilly Kolisko, 26 November 1928 (IWA).
248 Lilly Kolisko to Ita Wegman, 19 December 1928 (IWA).
249 LILLY KOLISKO: *Eugen Kolisko - Ein Lebensbild*, p. 372.
250 Lilly Kolisko to Ita Wegman, 26 February 1929 (IWA).
251 ELISABETH VREEDE: "Das Silber und der Mond. Experimentelle Studien von Lilly Kolisko". In: *Das Goetheanum*. VIII. Jahrgang, 1929, p. 230 - 231.
252 Lilly Kolisko to Ita Wegman, 21 November 1929 (IWA).
253 Lilly Kolisko to Ita Wegman, 26 February 1929 (IWA).
254 Lilly Kolisko to Dr Wachsmuth, 25 February 1929 (IWA).
255 Ita Wegman to Lilly Kolisko, 28 February 1929 (IWA).
256 LILLY KOLISKO: *Sternenwirken in Erdenstoffen III. Das Silber und der Mond. Experimentelle Studien aus dem Biologischen Institut am Goetheanum.* Stuttgart - The Hague - London 1929.
257 ELISABETH VREEDE: "Das Silber und der Mond. Experimentelle Studien von Lilly Kolisko". In: *Das Goetheanum*. VIII. Jahrgang, 1929, p. 230 - 231.
258 Lilly Kolisko to Ita Wegman, 23. May. 1929 (IWA).
259 LILLY KOLISKO: " Der Mond und das Pflanzenwachstum ". In: *Gäa Sophia, Volume IV. Landwirtschaft.* Yearbook of the Natural Science Section of the School of Spiritual Science at the Goetheanum. 1929, p. 94.
260 Ita Wegman to Lilly Kolisko, 27 October 1929 (IWA).
261 LILLY KOLISKO: "Über das Brot und das Quecksilber". In: *Natura.* 1929/30, p. 25 - 33.
262 RUDOLF HAUSCHKA: "Das Brot und die Erde". In: *Natura.* 1929/30, p. 7 - 18.
263 Cf. PETER SELG: *Rudolf Hauschka am Klinisch-Therapeutischen Institut in Arlesheim 1929 - 1941. Eine Dokumentation.* Arlesheim 2010, p. 44ff.
264 Ibid, p. 22ff.
265 RUDOLF HAUSCHKA: "Die Ernährung als kosmisch-irdisches Kräftespiel (Experimentelles zur Ernährung aus dem Versuchslaboratorium des Klinisch-Therapeut. Institutes, Arlesheim)". In: *Natura.* 1928/29, p. 340 - 364.
266 Lilly Kolisko to Ita Wegman, 21 November 1929 (IWA).
267 EUGEN KOLISKO AND LILLY KOLISKO: *Agriculture of Tomorrow.*

1946, p. 163.

268   Ibid, p. 182.

269   Cf. PETER SELG: *Rudolf Hauschka am Klinisch-Therapeutischen Institut in Arlesheim 1929 - 1941. Eine Dokumentation*, p. 94ff.

270   Ibid, p. 98.

271   Lilly Kolisko to Albert Steffen, 27 February 1930 (RSA and IWA).

272   PETER SELG: *Die Intentionen Ita Wegmans 1925 - 1943*, p. 200ff. ("Spiritualität am Abgrund. Ita Wegman und die Zivilisationsbedeutung der Michael-Schule").

273   Marie Steiner to Lilly Kolisko, 20 June 1928 (RSA).

274   Ita Wegman to Lilly Kolisko, 22 June 1928 (IWA).

275   Lilly Kolisko to Marie Steiner, 24 June 1928 (RSA).

276   PETER SELG: *Die Intentionen Ita Wegmans 1925 - 1943*, p. 239ff; JOHANNES KIERSCH: *Steiners individualisierte Esoterik einst und jetzt. Zur Entwicklung der Freien Hochschule für Geisteswissenschaft.* Dornach 2012, p. 301ff.

277   Minutes of the Executive Council meeting of 19 February 1930, Attachment "Brief an Frau Dr. Kolisko", 20 February 1930 (IWA).

278   Reply letters from Frau Kolisko to the Executive Council's letter dated 20 February 1930: Lilly Kolisko to Albert Steffen, 21 February 1930 (IWA).

279   Reply from Frau Kolisko to the letter from the Executive Council dated 20 February 1930: Lilly Kolisko to Ita Wegman, 21 February 1930 (IWA).

280   Lilly Kolisko to Albert Steffen, 27 February 1930 (RSA and IWA).

281   Lilly Kolisko to Guenther Wachsmuth, 23 March 1930 (GOEA). Also in: JOHANNES KIERSCH: *Zur Entwicklung der Freien Hochschule für Geisteswissenschaft. Die Erste Klasse,* p. 249 - 250.

282   Cf. LILLY KOLISKO: *Eugen Kolisko - Ein Lebensbild*, p. 246ff.

283   Marie Steiner to Albert Steffen, 17 March 1930 (GOA). In: JOHANNES KIERSCH: *Steiners individualisierte Esoterik einst und jetzt. Zur Entwicklung der Freien Hochschule für Geisteswissenschaft.* Dornach 2012, p. 103ff.

284   Ita Wegman to Albert Steffen, 27 June 1930 (IWA).

285   Ita Wegman to Eugen Kolisko, 2 January 1931 (IWA).

286   Lilly Kolisko to Albert Steffen, 27 February 1930 (RSA and

IWA).
281 Lilly Kolisko to Ita Wegman, 29 July 1931 (IWA).
287 LILLY KOLISKO: "Gedanken zu bevorstehenden Kalenderreformen". In: *Kalender Ostern 1931 – Ostern 1932*, published by the Mathematical and Astronomical Section at the Goetheanum. Dornach 1931, p. 60.
288 REINER PENTER: *Blick zurück nach vorn. Entwicklungswege und Verwandlungen des Anthroposophischen Klinisch-Therapeutischen Impulses in Unterlengenhardt.* Bad Liebenzell-Unterlengenhardt 2004, p. 38.
289 LILLY KOLISKO: *Eugen Kolisko - Ein Lebensbild*, p. 242ff.
290 From his mother, Amalie Kolisko, who died in 1927.
291 Lilly Kolisko to Albert Steffen, 27 February 1930 (RSA and IWA).
281 Lilly Kolisko to Ita Wegman, 29 July 1931 (IWA).
292 Ita Wegman to Lilly Kolisko, 7 August 1931 (IWA).
293 Lilly Kolisko to Ita Wegman, 14 August 1931 (IWA).
294 ELISABETH VREEDE: "Bericht über die Astronomische Tagung vom 26 – 28 August 1931". In: *Mathematisch-Astronomische Sektion.* Dornach 1931, p. 1 (GoA).
295 Ibid, p. 2.
296 LILLY KOLISKO: "Gedanken zu bevorstehenden Kalenderreformen In: Kalender Ostern *1931* - Ostern *1932*, In: *Mathematisch-Astronomische Sektion.* Dornach 1931, p. 57 - 58.
297 Ibid, p. 60 - 62.
298 LILLY KOLISKO: *Mitteilungen des Biologischen Institutes am Goetheanum No. 1*, p. 42.
299 EUGEN KOLISKO AND LILLY KOLISKO: *Agriculture of Tomorrow* p. 56.
300 Ibid, p. 57.
301 LILLY KOLISKO: *Sternenwirken in Erdenstoffen. Der Jupiter und das Zinn. Experimentelle Studien aus dem Biologischen Institut am Goetheanum.* Stuttgart 1932.
302 LILLY KOLISKO: *Workings of the Stars in Earthly Substances IV. Jupiter and Tin. Experimental Studies from the Biological Institute of the Goetheanum.* Stuttgart 1932.
303 LILLY KOLISKO: *Sternenwirken in Erdenstoffen. Der Jupiter und das Zinn*, p. 12

304 Ita Wegman to Eugen Kolisko, 14 December 1928 (IWA).

305 RUDOLF STEINER: *Initiationswissenschaft und Sternenerkenntnis.* GA 228. Dornach 2002, p. 99.

306 Invitation to Lectures and Exhibition of the Work of L. Kolisko, Biological Institute of the Goetheanum, Stuttgart, November 27th to December 4th , 1932, *Anthroposophical Society in Great Britain in Conjunction with the Anthroposophical Agricultural Foundation.*

307 Handwritten notes by Gladys Knapp (C. R. / S. T.).

308 LILLY KOLISKO: *Physiologischer Nachweis der Wirksamkeit kleinster Entitäten.* Stuttgart 1932.

309 LILLY KOLISKO: *Der Mond und das Pflanzenwachstum.* Stuttgart 1933.

310 Ibid, p. 34.

311 Walter Johannes Stein to Ita Wegman, 8 May 1933 (IWA). On these events, see the monograph by UWE WERNER: *Anthroposophen in der Zeit des Nationalsozialismus (1933 - 1945).* Munich 1999.

312 LILLY KOLISKO: *Eugen Kolisko - Ein Lebensbild*, p. 341.

313 Ibid, p. 373.

314 On the observation of Eugen Kolisko by the Nazi authorities, see PETER SELG, SUSANNE H. GROSS, MATTHIAS MOCHNER: *Anthroposophische Medizin, Pharmazie und Heilpädagogik im Nationalsozialismus 1933 - 1945. Volume 1: Anthroposophie und Nationalsozialismus. Die anthroposophische Ärzteschaft.* Basel 2024. Chap. 7.2.1.

315 Cf. PETER SELG: *Erzwungene Schließung. Die Ansprachen der Stuttgarter Lehrer zum Ende der Waldorfschule im deutschen Faschismus (1938).* Arlesheim 2019, p. 286ff.

316 LILLY KOLISKO: *Eugen Kolisko - Ein Lebensbild*, p. 373.

317 Cf. PETER SELG: "Eugen Kolisko". In: *Anfänge anthroposophischer Heilkunst*, p. 167f.

318 Ita Wegman to Lilly Kolisko, 12 January 1932 (IWA).

319 Lilly Kolisko to Madeleine van Deventer, 16 December 1932 (IWA).

320 Ernst Aisenpreis to Lilly Kolisko, 31 August 1934, in: *Korrespondenz der Anthroposophischen Arbeitsgemeinschaft, Deutschland.*

321 Ibid.
322 PAUL GIMMI: "Oktober 1934". In: *Korrespondenz der Anthroposophischen Arbeitsgemeinschaft.*
323 LILLY KOLISKO: *Mitteilungen des Biologischen Instituts am Goetheanum.* No. 1, St John's 1934; No. 2, Christmas 1934; No. 3 and No. 4 were published in 1935.
324 LILLY KOLISKO: *Kristall-Gestaltungskräfte I. Experimentelle Studien aus dem Biologischen Institut am Goetheanum.* Stuttgart 1936.
325 REINER PENTER: *Blick zurück nach vorn Entwicklungswege und Verwandlungen des Anthroposophischen Klinisch-Therapeutischen Impulses in Unterlengenhardt.* Bad Liebenzell-Unterlengenhardt, 2004, p. 16 and 21.
326 Cf. PETER SELG, HEIKE S. GROSS, MATTHIAS MOCHNER *Anthroposophische Medizin, Pharmazie und Heilpädagogik im Nationalsozialismus 1933 - 1945. Volume 1: Anthroposophie und Nationalsozialismus. Die anthroposophische Ärzteschaft.* Chap. 7.2.1.
327 REINER PENTER: *Blick zurück nach vorn. Entwicklungswege und Verwandlungen des Anthroposophischen Klinisch-Therapeutischen Impulses in Unterlengenhardt,* p. 21 - 22.
328 Helene von Grunelius to Margarete Bockholt, 9 January 1935 (IWA).
329 EUGEN KOLISKO AND LILLY KOLISKO: *Agriculture of Tomorrow,* p. 172.
330 Cf. J. E. ZEYLMANS VAN EMMICHOVEN: *Wer war Ita Wegman. Eine Dokumentation. Volume 3.*
331 LILLY KOLISKO: *Eugen Kolisko - Ein Lebensbild,* p. 378 - 391.
332 J. E. ZEYLMANS VAN EMMICHOVEN: *Wer war Ita Wegman. Eine Dokumentation. Volume 3,* p. 333ff; and PETER SELG (ed.): *Widerspruch. Ungehörte und verdrängte Stimmen gegen die Dornacher Ausschlüsse des Jahres 1935.* Arlesheim, 2028, p. 126ff.
333 LILLY KOLISKO: *Eugen Kolisko – Ein Lebensbild,* p. 391 - 392.
334 Lilly Kolisko to Ita Wegman, 19 July 1935 (IWA).
335 Ita Wegman to George Adams Kaufmann, 19 August 1935 (IWA). Cf. PETER SELG: *Rudolf Steiner und die Freie Hochschule für Geisteswissenschaft.* Arlesheim 2008, p. 79. See in detail on

Ita Wegman's handling of the Class lessons after 14 April 1935: PETER SELG: *Die Intentionen Ita Wegmans 1925 - 1943*, p. 265ff.

336 Ita Wegman to Lilly Kolisko, 10 August 1935 (IWA).

337 ELISABETH VREEDE to Willem Zeylmans van Emmichoven, 20 October 1935 (IWA).

338 Lilly Kolisko to Ita Wegman, 18 June 1936 (IWA).

339 Ita Wegman to Lilly Kolisko, 6 August 1936 (IWA).

340 ELISABETH VREEDE: Postcard to Haus Vreede, 19 June 1936 (IWA).

341 LILLY KOLISKO: *Mitteilungen des Biologischen Instituts am Goetheanum, No. 5 (Sonnenfinsternis 19 November 1936 in Brussa, Turkey)*. Stuttgart 1936, p. 12.

342 Elisabeth Vreede to the Friends in Arlesheim, 17 and 21 June 1936 (IWA).

343 LILLY KOLISKO: *Mitteilungen des Biologischen Instituts am Goetheanum, No. 5 (Sonnenfinsternis 19 November 1936 in Brussa, Turkey)*. Stuttgart 1936. and LILLY KOLISKO: *Gold and the Sun. An Account of Experiments Conducted in Connection with the Total Eclipse of the Sun of 19 June 1936*. London 1936.

344 Cf. PETER Selg: *Elisabeth Vreede. 1879 - 1943*. Arlesheim 2009, p. 208ff.

345 ELISABETH VREEDE: "The Total Eclipse of the Sun 19 June 1936". In: *The Present Age*. Vol. I, No. 10, 1936, p. 20 ff.; reprinted in: MADELEINE P. VAN DEVENTER, ELISABETH KNOTTENBELT (eds.): *Elisabeth Vreede: Ein Lebensbild*. Arlesheim 1976, p. 92 - 99; and PETER SELG: *Elisabeth Vreede. 1879 - 1943*, p. 210 - 219. See also the facsimile and transcription of Vreede's travel postcard from Bursa dated 19. June 1936 with her experiences on the mountain. Ibid, p. 215.

346 Eugen Kolisko to Ita Wegman, 18 November 1936 (IWA).

347 LILLY KOLISKO: *Eugen Kolisko - Ein Lebensbild*, p. 395.

348 Madeleine van Deventer to Ita Wegman, 30 March 1936 (IWA).

349 Ita Wegman to Fried Geuter, 27 May 1936 (IWA).

350 Summer School on "Anthroposophy and the Free Life of Spirit", Normal College, Bangor 4 August to 12 August 1936. LILLY KOLISKO: *Lantern Lecture: Experimental Research on the Solar Eclipse on June 19, 6 August at 8.30 p.m.* (IWA)

351   See LILLY KOLISKO: *Eugen Kolisko - Ein Lebensbild*, p. 396 - 397.

352   Syllabus of Lectures to be held during the Autumn Season 1936 at Rudolf Steiner House, 35 Park Road, N.W. I, Anthroposophical Society in Great Britain (IWA).

353   Eugen Kolisko to Ita Wegman, 18 November 1936 (IWA).

354   Lilly Kolisko to Ita Wegman, 18 June 1937 (IWA).

355   *The Modern Mystic and Monthly Science Review. A Monthly Journal devoted to the Study of Mysticism and Occult Sciences.*

356   LILLY KOLISKO: "Prelude to Scientific Research". In: *The Modern Mystic*. Vol. 1, No. 5, June 1937.

357   LILLY KOLISKO: "The Influence of the Moon". Ibid, Vol. 1, No. 6, July 1937.

358   LILLY KOLISKO: "Moon and Silver". Ibid, Vol. 1, No. 7, August 1937; "Gold and the Sun", Ibid. Vol. 1, No. 8, September 1937, "Mars and Iron, Jupiter and Tin, Saturn and Lead". Ibid. Vol. 1, No. 9, October 1937.

359   LILLY KOLISKO: "Is Matter really Material? Research into the Influence of the Infinitesimal". Ibid. Vol. 1, No. 10, November 1937.

360   Ita Wegman to Eugen Kolisko, 14 December 1928 (IWA).

361   Eugen Kolisko to Walter J. Stein, 17 August 1937 (IWA).

362   "Appeal, Biological Research Institute for Investigating COSMIC Influences on Earthly Substances and Living Organisms in Connection with the School of Spiritual Science". In: *The Modern Mystic.* Vol. 1, No. 9, October 1937.

363   ELEANOR C. MERRY: "A New Beginning". In: *The Modern Mystic.* Vol 1, No. 11, December 1937.

364   George Adams Kaufmann to Ita Wegman, 27 June 1937 (IWA).

365   Ita Wegman to George Adams Kaufmann, 2 July 1937 (IWA).

366   George Adams Kaufmann to Ita Wegman, 4 July 1937 (IWA).

367   Note (RSA).

368   "Versuche an verschiedenen Orten der Erde". In: *Besprechungen im Stuttgarter Ärztekreis*. November 1949.

369   LILLY KOLISKO: "Preliminary Report of my Stay in Calcutta". In: *The Present Age*. Vol. 3, No. 5, May 1938, p. 12 - 13.

370   Walter Johannes Stein to Ita Wegman, 6 May 1937 (IWA).

371   "Versuche an verschiedenen Orten der Erde". In: *Besprechungen im Stuttgarter Ärztekreis*. November 1949.

372   WALTER JOHANNES STEIN: "Mrs Kolisko's Journey to India".

In: *The Present Age.* Vol. 3, No. 2, February 1938, p. 60 - 62.

373 LILLY KOLISKO: "Preliminary Report of my Stay in Calcutta". In: *The Present Age.* Vol. 3, No. 5, May 1938, p. 12 - 13.

374 EUGEN KOLISKO AND LILLY KOLISKO: *Agriculture of Tomorrow*, p. 16 - 17.

375 Lilly Kolisko described in 1949 how she carried out her potentiations: "Dr. Steiner once demonstrated how to shake: quite quickly, but lightly and gently back and forth, essentially from the wrist, quite evenly, without interruption. The sound will be the best way to perceive how one shakes." In: *Besprechungen mit Frau Kolisko vom 3. bis 5. November 1949 in Schwäbisch Gmünd.* Minutes. Weleda Archive, Arlesheim.

376 LILLY KOLISKO: *Physiologischer und physikalischer Nachweis der Wirksamkeit kleinster Entitäten 1923 - 1959*, p. 271 - 272.

377 Eugen Kolisko to Ita Wegman, 2 June 1938 (IWA).

378 LILLY KOLISKO: *Eugen Kolisko – Ein Lebensbild*, p. 397 - 398.

379 Maria Röschl to Ita Wegman, 9 December 1939 (IWA).

380 KARL KÖNIG: Diary note, 9 December 1938 (KKA).

381 LILLY KOLISKO: *Eugen Kolisko - Ein Lebensbild*, p. 397. However, Karl König gave a completely different picture of Eugen Kolisko's experiences in the USA: "We sat in a small pub in London and he talked about America. About the people he met there, about his lectures, which were all poorly attended, and he could only with difficulty overcome the deep disappointment that had overcome him. He looked tired and ill, but he courageously developed ideas for the future." In: PETER SELG: "Eugen Kolisko". In: *Anfänge anthroposophischer Heilkunst.* Dornach 2000, p. 172.

382 Ibid, p. 397.

383 Vera Piper to Ita Wegman, 7 September 1939 (IWA).

384 LILLY KOLISKO: *Eugen Kolisko – Ein Lebensbild*, p. 397 - 399.

385 MAGDA MEYER: Memories conveyed in a conversation, S. T.

386 JOHANNES EMANUEL ZEYLMANS VAN EMMICHOVEN: *Willem Zeylmans van Emmichoven: Ein Pionier der Anthroposophie.* Arlesheim 1979, p. 231.

387 Maria Röschl to Ita Wegman, 9 December 1939 (IWA).

388 Walter Johannes Stein to Ita Wegman, 11 December 1939 (IWA, original in English).

389   Ita Wegman to Lilly Kolisko, 12 December 1939 (IWA, original in English).

390   In: EBERHARD SCHICKLER, JULIA MELLINGER, JÜRGEN VON GRONE: *Eugen Kolisko. Bilder aus seinem Leben und Wirken.* Stuttgart 1940.

391   LILLY KOLISKO: *Eugen Kolisko - Ein Lebensbild, p.* 419 - 420.

392   Walter Johannes Stein to Ita Wegman, 8 April 1940 (IWA).

393   EUGEN KOLISKO AND LILLY KOLISKO: *Agriculture of Tomorrow.* Note, p. vi.

394   LILLY KOLISKO: "Astro-Biological Calendar" for February & March & March and April / May / June / July / August / September / October / November and December / January. In: *The Modern Mystic.* Vol. 3, 1939. No. 1 - 11 and January 1940 No . 12.

395   LILLY KOLISKO: "Moon and Silver", February 1940, in: *Tomorrow*, p. 360 - 363; "Gold and Sun", March 1940, ibid. p. 398 - 400; "Mars and Iron, Jupiter and Tin, Saturn and Lead", April 1940, ibid. p. 454 - 456.

396   Cf. PETER SELG, SUSANNE H. GROSS, MATTHIAS MOCHNER: *Anthroposophische Medizin, Pharmazie und Heilpädagogik im Nationalsozialismus 1933 - 1945. Volume 1: Anthroposophie und Nationalsozialismus. Die anthroposophische Ärzteschaft.* Chap. 7.2.2.

397   Ita Wegman to Walter Johannes Stein, 31 March 1940 (IWA, original in English).

398   Walter Johannes Stein to Ita Wegman, 8 April 1940 (IWA).

399   Maria Schmidler to Lilly Kolisko, 12 November 1940 (C. R. / S. T.).

400   Norbert Glas to Ita Wegman, 12 April 1941 (IWA, original in English).

401   EUGEN KOLISKO AND LILLY KOLISKO: *Agriculture of Tomorrow.* Edge 1946.

402   EUGEN KOLISKO AND LILLY KOLISKO: *Agriculture of Tomorrow, Edge 1946* p. vi.

403   ILSE KETZEL: handwritten diary, library of the Waldorf School in Buckfastleigh, Devon, England. (Original in English)

404   GEORG MEYER: school reports from 1941 from Wynstones Waldorf School, Georg Meyer estate.

405   From conversations with Marguerite A. Wood, Soili Turunen.

406   From conversations with Georg and Magda Meyer, Soili Turunen.

407　EUGEN KOLISKO AND LILLY KOLISKO: *Capillary Dynamolysis. Advance Print of some chapters from the second part of the book "Agriculture of Tomorrow".* Wynstones 1943.

408　EUGEN KOLISKO: *Three Fundamental Problems of the Anthroposophical Knowledge of Man; I. The Bodily Foundation of Thinking; II. The Bodily Foundation of Feeling; III. The Bodily Foundation of Human Will; IV. Man's Connection with the Whole Universe.* Wynstones 1944. Reprint see www.koliskoarchive.com, in: *The Threefold Human Organism.*

409　EUGEN KOLISKO: *Three Fundamental Problems of the Anthroposophical Knowledge of Man; I. The Bodily Foundation of Thinking.* In Editorial, p. I.

410　EUGEN KOLISKO: *Nutrition I., II., III.*, Wynstones 1944; *Zoology for Everybody 1 - 8*, Wynstones 1944 - 45, *Geology 1 – 5*, Wynstones 1945. Reprint see www.koliskoarchive.com.

411　G. A. M. KNAPP: *Darkness and Light, Expressed in diagonal shading.* Bournemouth 1982.

412　From conversations with Marguerite A. Wood, Soili Turunen.

413　Later published in: LILLY KOLISKO: *Spirit in Matter. A Scientific Answer to the Bishop's Queries.* Edge 1948.

414　Karl König received the news of Ita Wegman's death (4 March 1943) the same evening by telegram from Arlesheim. A memorial service was held at Camphill on the day of Wegman's death. Cf. PETER SELG: *Ita Wegman and Karl König. Eine biographische Dokumentation.* Dornach 2007, p. 86.

415　Lilly Kolisko to Gladys Knapp, 29 December 1944 (C. R. / S. T., original in English).

416　GRACE KNAPP: Handwritten manuscript: "Building the Hut" The Kolisko Archives, 1978 (C. R. / S. T., original in English).

417　Lilly Kolisko to Grace Knapp, 27 May 1945 (C. R. / S. T., original in English).

418　Lilly Kolisko to Grace Knapp, 1 June 1945 (C. R. / S. T., original in English).

419　Lilly Kolisko to Grace Knapp, 9 June 1945 (C. R. / S. T., original in English).

420　Lilly Kolisko to Grace Knapp, 27 June 1945 (C. R. / S. T., original in English).

421　Draft for membership of the Kolisko Archive (C. R. / S. T., original in English).

422   Ibid.

422   ILSE KETZEL: handwritten diary. (original in English).

423   Lilly Kolisko to V. Gorsky, 30 April 1947 (RSA, C. R. / S. T., original in English).

425   LILLY KOLISKO: *Spirit in Matter*. Preface, p. IV.

426   In: EUGEN KOLISKO AND LILLY KOLISKO: *The Agriculture of Tomorrow*. 1953. afterword for the English edition, May 1946, p. 382 - 384.

427   EUGEN KOLISKO AND LILLY KOLISKO: *Die Landwirtschaft der Zukunft*. Foreword to the German edition, p. I.

428   Lilly Kolisko to Gladys Knapp, 13 June 1946 (C. R. / S. T., (C. R. / S. T., original in English).

429   Summer Conference, "Lessons of the War", 29 July to 5 August 1946, Wynstones Waldorf School.

429   Lilly Kolisko to V. Gorsky, 6 March 1947 (original in English)

430   Lilly Kolisko to V. Gorsky, 30 April 1947 (RSA, original in English).

432   DR OSKAR SCHMIEDEL: "E. and L. Kolisko, Agriculture of Tomorrow". In: *Weleda Nachrichten*. No. 44, Arlesheim 1948, p. 1 - 7.

433   LILLY KOLISKO: *Gold and the Sun. An Account of Experiments Conducted in Connection with the Total Eclipse of the Sun of 20 May, 1947*. Edge 1947, p. 7.

434   LILLY KOLISKO: *Gold and the Sun. An Account of Experiments Conducted in Connection with the Total Eclipse of the Sun of 20th May 1947*. Edge 1947.

435   LILLY KOLISKO: *Spirit in Matter. A Scientist's Answer to the Bishop's Queries*. Edge 1948, p. IV.

436   Ibid, p. 21.

437   Ibid, p. 27.

438   M. E. BRUCE: *Common Sense Compost Making, by the Q. R. (Quick Return) Composting Solutions*. London 1946.

439   LILLY KOLISKO: *Agriculture of Tomorrow preparations*. Agriculture of Tomorrow Publications I, Edge 1949.

440   Ibid, p. 9.

441   Ibid, p. 14.

442   EUGEN KOLISKO AND LILLY KOLISKO: *Die Landwirtschaft der*

Zukunft. Part III, Rudolf Steiner's advice for farmers, p. 233 - 341, 1953.

443 LILLY KOLISKO: *Agriculture of Tomorrow preparations*, p. 9.

444 IRIS BLEISCH: "Vorträge von Lilly Kolisko". In: *Beiträge zur Erweiterung der Heilkunst.* No. 3 /1952, p 140 - 142.

445 Fried Geuter to Karl König, 17 September 1953 (KKA).

446 JÜRGEN V. GRONE: "Lili Kolisko in Stuttgart". In: *Beiträge zur Erweiterung der Heilkunst.* 6 /1949, p 409 - 411.

447 *Besprechungen mit Frau Kolisko vom 3 bis 5 November 1949 in Schwäbisch Gmünd.* Protokoll. Weleda Archiv, Arlesheim.

448 Ibid, p 1.

449 Ibid, p 4.

450 Ibid, p. 7.

451 Ibid, p. 8.

452 THEODOR SCHWENK: "Erinnerungen zur Steigbildmethode (Kapillar-dynamolyse). Im Gedenken an den 95. Geburtstag von Lilly Kolisko am 31. August 1984". In: *Mitteilungen aus der anthroposophischen Arbeit in Deutschland.* Issue 3 /1984, p. 199 - 200.

453 BARBARA SCHAEFFER: *La Motta in Erzählungen 1938 - 2006*, p. 193.

454 The original pictures have been preserved (C. R. / S. T.).

455 IRIS BLEISCH: "Vorträge von Lilly Kolisko". In: *Beiträge zur Erweiterung der Heilkunst.* No. 3 /1952, p. 140 - 142.

456 See more details in: LILLY KOLISKO: *Sternenwirken in Erdenstoffen. Saturn und Blei. Ein Versuch, die Phänomene der Chemie, Astronomie und Physiologie zusammen zu schauen.* Edge 1952.

457 IRIS BLEISCH: "Vorträge von Lilly Kolisko". In: *Beiträge zur Erweiterung der Heilkunst.* No. 3 /1952, p. 140 - 142.

458 GISBERT HUSEMANN: "Lili Kolisko, Werk und Wesen", p. 53.

459 CHRISTIAN LAHUSEN: "Die Weihnachtszeit erlebt im naturwissenschaftlichen Experiment, Vortrag von Frau Lilly Kolisko, Edge (England) von 6. Januar 1952 in Stuttgart". In: *Beiträge zur Erweiterung der Heilkunst.* 3 /1952, p. 142 - 143.

460 Knut Clunies- Ross, 4. June 1923 – 13. April 2010.

461 Foreword in: LILLY KOLISKO: *Sternenwirken in Erdenstoffen, Saturn und Blei.* Edge 1952.

462 Ibid, p. 161 - 172.

463   LILLY KOLISKO: "Saturn-Chronos". In: Norbert Glas: *Lichtvolles Alter*. Stuttgart 1956.

464   *Hippocrates. Zeitschrift für praktische Heilkunde vom Organ des Kneippärztebundes und Organ des Zentralverbandes der Ärzte-Gesellschaft für Naturheilverfahren.* Edited by Th. Dobler and Heinrich K. Kunstmann.

465   Eberhard Schickler to Karl König, 12 May 1952 (KKA).

466   Eberhard Schickler to Karl König, 2 June 1952 (KKA).

467   Eberhard Schickler to Karl König, 5 July 1952 (KKA).

468   LILLY KOLISKO: "Aus Forschung und Wissenschaft: Die Kapillar-Dynamolysis. Eine spezifische Methode zum Studium der Gestaltungskräfte von unorganischen und organischen Substanzen. Die Anwendung dieser Methode in der Medizin, Lebensmittelkunde und Landwirtschaft". In: *Hippokrates. Zeitschrift für praktische Heilkunst.* 24. Jahrgang, Heft 5 /1953, p. 130 - 135.

469   LILLY KOLISKO: *Eugen Kolisko – Ein Lebensbild*, p. 444.

470   EUGEN KOLISKO AND LILLY KOLISKO: *Die Landwirtschaft der Zukunft*. Schaffhausen 1953.

471   Fried Geuter to Karl König, 17 September 1953 (KKA).

472   GLADYS ANNIE M. KNAPP: "Spirit in Matter. The Research work of Lilly Kolisko". In: *Anthroposophical Quarterly.* London, Vol. 22, Number 1, Spring 1977, p. 12; and in: *Journal for Anthroposophy.* London, Number 26, Autumn 1977, p. 13.

473   Lilly Kolisko to Norbert Glas, 6 April 1958 (IWA).

474   RUDI LISSAU in Lilly Kolisko Quarter Century Conference, Stroud 2001.

475   Obituary, © Kolisko Archive, London.

476   LILLY KOLISKO: *Eugen Kolisko – Ein Lebensbild*, p. 5.

477   GISBERT HUSEMANN in the blurb of the book cover of LILLY KOLISKO: *Physiologischer und physikalischer Nachweis der Wirksamkeit kleinster Entitäten, 1923 - 1959*. Stuttgart 1959.

478   LILLY KOLISKO: *Physiologischer und physikalischer Nachweis der Wirksamkeit kleinster Entitäten, 1923 - 1959*, Stuttgart 1959.

479   Eberhard Schickler to Karl König, 12 June 1958 (KKA).

480   Gisbert Husemann in the blurb of the book cover of LILLY KOLISKO: *Physiologischer und physikalischer Nachweis der Wirksamkeit kleinster Entitäten, 1923 - 1959*.

481   Michael Schad to Karl König, 4 October 1960 (KKA).

482   LILLY KOLISKO: "Die totale Sonnenfinsternis vom 15. Februar 1961, beobachtet in Bordighera (Norditalien), vorläufiger Kurzbericht", p. 169 - 175; ADALBERT STIFTER: "Anblick an der Welt (Sonnenfinsternis am 8. Juli 1842", p. 176 - 179; RUDOLF STEINER: "Das Wesen der Sonnenfinsternis" (From: *Menschenfragen und Weltenantworten*. Dornach, 1922), p. 179 - 180. In: *Beiträge zu einer Erweiterung der Heilkunst nach geistes-wissenschaftlichen Erkenntnissen*. Stuttgart 1961, issue no. 5, September / October 1961.

483   LILLY KOLISKO, ADALBERT STIFTER, RUDOLF STEINER: *Die totale Sonnenfnsternis im Experiment, als Erlebnis und ihr Wesen*. Stuttgart 1961. In English: LILLY KOLISKO, ADALBERT STIFTER, RUDOLF STEINER: *The Sun Eclipse in Experiment, as Experience, its Nature*. BOURNEMOUTH 1978.

484   P.E.Schiller to Dr Schwenk, 10 March 1961, from the estate of Paul Eugen Schiller.

485   Lilly Kolisko to Gladys Knapp, 31 August 1961 (C. R. / S. T., original in English).

486   LILLY KOLISKO: *Eugen Kolisko – Ein Lebensbild*, p. 5.

487   Lilly Kolisko to Gladys Knapp, 19 June 1961 (C. R. / S. T., original in English).

488   Lilly Kolisko to Gladys Knapp, 10 July 1961 (C. R. / S. T., original in English).

489   Margarete Bockholt to Hilma Walter, 20 December 1961 (IWA).

490   Gisbert Husemann to Karl König, 8 March 1966 (KKA).

491   Karl König to Gisbert Husemann, 22 March 1966 (KKA).

492   LILLY KOLISKO: "Rudolf Steiner über Potenzierung und Hochpotenzen". From a letter to Hans Krüger, 6 November 1961. in: *Der Merkurstab*. 5 /1994, p. 507; and in: *Weleda Korrespondenzblätter für Ärzte*. No. 138, September 1994, p. 86 - 87.

493   Ibid.

494   Gisbert Husemann to Nobert Glas and Karl König, undated. (before Christmas 1963) (KKA).

495   RUDOLF HAUSCHKA: *Wetterleuchten der Zeitenwende*. Bad Boll 2007, p. 149 - 150.

496   Lilly Kolisko to Gladys Knapp, 8 December 1962 (C. R. / S. T., original in English).

497 Lilly Kolisko: Notebook (RSA).

498 Lilly Kolisko to Gladys Knapp, 4 January 1966 (C. R. / S. T., original in English).

499 Jerôme Noever de Brauw to Soili Turunen, 20 January 2020 (S. T.).

500 WILLY SUCHER, 21. 8. 1902 - 21. 5. 1985. Cf. a. www.biographien. kulturimpuls.org and www.astrosophy.com.

501 Lilly Kolisko to Gladys Knapp, 8 December 1962 (C. R. / S. T., original in English).

502 Lilly Kolisko to Gladys Knapp, 9 January 1963 (C. R. / S. T., original in English).

503 Lilly Kolisko to Gladys Knapp, 5 July 1965 (C. R. / S. T., original in English).

504 Lilly Kolisko to Gladys Knapp, 1 January 1969 (C. R. / S. T., original in English).

505 Lilly Kolisko to Gladys Knapp, 6 December 1969 (C. R. / S. T., original in English).

506 Original drawings by GLADYS KNAPP: "Drawings of Gladys Knapp out of her work with Lilly Kolisko". Biodynamic Association, Stroud.

507 Ibid. Text on the drawing of Chelidonium majus: "Dr Steiner gave me the task of showing the connections between the minerals, plants and planets." Frau L. Kolisko.

508 GISBERT HUSEMANN: "Lili Kolisko - Werk und Wesen", p. 53.

509 Ibid, p. 53.

510 THOMAS JURRIAANSE: "Lili Kolisko". In: *Mededelingen voor Leden van de Antroposofische Vereniging in Nederland.* 1977, p. 17.

511 WILHELM PELIKAN: "Zum Lebenswerk von Lilly Kolisko". In: *Das Goetheanum.* No. 13 /1977, p. 100.

512 LILLY KOLISKO: *Spirit in Matter*, p. 37.

ITA WEGMAN INSTITUTE
FOR BASIC ANTHROPOSOPHICAL RESEARCH

At the Ita Wegman Institute for Basic Anthroposophical Research, the anthroposophical spiritual science developed by Dr phil. Rudolf Steiner (1861 - 1925) in written and lecture form is analysed in terms of the history of ideas, with a biographical emphasis and in the context of the scientific and social history of the 19th and 20th centuries.

The Institute maintains several publicly accessible work archives based on the estates of Rudolf Steiner's pioneering colleagues, particularly in the fields of medicine, curative education and pedagogy.

The work of the Ita Wegman Institute is supported by various foundations - primarily the Software AG Foundation (Darmstadt) - as well as an international circle of friends and sponsors.

Pfeffinger Weg 1A - CH 4144 Arlesheim - Switzerland
Management: Prof. Dr P. Selg
www.wegmaninstitut.ch - E-Mailsekretariat@wegmaninstitut.ch

www.ingramcontent.com/pod-product-compliance
Lightning Source LLC
Chambersburg PA
CBHW051441270326
41932CB00025B/3390